Institution of Civil Engineers

publishing

Risk Assessments Questions and Answers

A practical approach

Second edition

Pat Perry MCIEH, MIIRM, FRSH, MIOSH

Published by ICE Publishing, One Great George Street, Westminster, London SW1P 3AA.

Full details of ICE Publishing sales representatives and distributors can be found at: www.icebookshop.com/bookshop_contact.asp

Other titles by ICE Publishing:

Health and Safety, Questions and Answers. A practical approach, Second edition.
P. Perry. ISBN 978-0-7277-6074-6

CDM 2015, Questions and Answers. A practical approach.
P. Perry. ISBN 978-0-7277-6032-6

ICE Manual of Health and Safety in Construction, Second edition.
C. McAleenan and D. Oloke (eds). ISBN 978-0-7277-6010-4

www.icebookshop.com

A catalogue record for this book is available from the British Library

ISBN 978-0-7277-6076-0

© Thomas Telford Limited 2017

ICE Publishing is a division of Thomas Telford Ltd, a wholly-owned subsidiary of the Institution of Civil Engineers (ICE).

Commissioning Editor: Amber Thomas
Development Editor: Maria Inês Pinheiro
Production Editor: Abigail Neale
Market Development Executive: Elizabeth Hobson

Typeset by Academic + Technical, Bristol
Index created by Nigel D'Auvergne
Printed and bound by Bell & Bain Limited, Glasgow, UK

MIX
Paper from
responsible sources
FSC
www.fsc.org
FSC® C007785

Risk Assessments
Questions and Answers

A practical approach

Other titles in the Questions and Answers series:

Health and Safety, Questions and Answers. A practical approach.
Second edition (2016)
P. Perry. ISBN 978-0-7277-6074-6

CDM 2015, Questions and Answers. A practical approach. (2015)
P. Perry. ISBN 978-0-7277-6032-6

Contents

Acknowledgements

I would like to thank my family for all their support and patience throughout the numerous days of writing and typing the manuscript for this book.

About the author

Pat Perry, MCIEH, MIOSH, FRSH, MIIRM, qualified as an Environmental Health Officer in 1978 and spent the early years of her career in local government enforcing environmental health regulations, particularly health and safety law, which became her passion. She has extensive knowledge of the sector and has served on various working parties on both health and safety and food safety issues. Pat also regularly contributes to professional journals and appears frequently as a speaker at related conferences and seminars.

Pat set up her own environmental health consultancy, Perry Scott Nash Associates Ltd (PSN), in 1988, and fulfilled her vision of a creating a customer-focused provider of consultancy services to the commercial, hospitality and retail sectors.

The consultancy grew considerably over the years and provided advice to a wide range of high-profile clients across a variety of market sectors. In 2013, PSN joined food supply chain risk management Group Acoura. As an industry leader in its chosen field Acoura has been able to provide support and expertise that has enabled PSN to realise improved growth while also providing customers with the broad range of services and solutions they require.

Hospitality, leisure and retail have become the consultancy's main sectors of specialisation, and PSN's team is particularly well versed in delivering construction, design and management (CDM) and environmental health services to clients operating in these areas. The team works on projects worth from a few hundred thousand pounds to many millions, with new public housing developments and major department store refits and refurbishments among the types of developments it assists with.

Now known as the specialist division of its parent company, Acoura, PSN continues to build strong links with enforcing agencies, while also providing pragmatic, innovative solutions to compliance issues across a wide range of industries.

List of abbreviations

ACOP	approved code of practice
CAD	computer-aided design
CDM	construction, design and management
CDM 2015	Construction (Design and Management) Regulations 2015
COSHH	Control of Substances Hazardous to Health Regulations 2002
DSE	display screen equipment
DSEAR	Dangerous Substances and Explosive Atmospheres Regulations 2002
EU	European Union
FFI	fees for intervention (scheme)
HSE	Health and Safety Executive
MEL	maximum exposure limit
MHSW	Management of Health and Safety at Work (Regulations 1999)
OEL	occupational exposure limit
OES	occupational exposure standard
PACE	Police and Criminal Evidence Act 1984
PPE	personal protective equipment
RSI	repetitive strain injury
STEL	short-term exposure limit
SWL	safe working load
TWA	time-weighted average
WEL	workplace exposure limit
WRULD	work-related upper limb disorder

Risk Assessments: Questions and Answers
ISBN 978-0-7277-6076-0

ICE Publishing: All rights reserved
http://dx.doi.org/10.1680/raqa.60760.001

Chapter 1
Introduction

What does the term 'risk assessment' mean?

Quite literally, it means an assessment of the likelihood of something going wrong or affecting somebody or something in a way which could cause them harm, or damage property, etc.

Risk assessment is a logical approach to reviewing the dangers in a job, activity or event and determining the consequences. The consequences may be acceptable to the assessor or they may not. If the dangers of a job or activity are considered too high then the activity should not be undertaken unless control measures are put in place (i.e. steps are taken to reduce the risk of harm or injury to an acceptable level).

Why is risk assessment so important in respect of health and safety?

Employees and others undertake a vast array of jobs and activities in their working environment and it is important that employers protect their safety at all times. In order to know what the hazards and risks are in respect of the job activity the employer will need to assess the job so as to be able to identify them.

Employers need a systematic way of being able to review and record the identified hazards associated with a job, and the risk assessment process provides this.

Is risk assessment something new?

No. There has been an implied requirement to carry out risk assessment in the Health and Safety at Work etc. Act 1974, as Section 2 requires employers to do things 'as far as is reasonably practicable'. In order to assess what is reasonably practicable it has always been necessary to review practices, procedures, tasks and jobs to find out what is being done and what can be done to ensure the safety of individuals.

The first formal requirement for risk assessment under legislation was probably in the Control of Substances Hazardous to Health Regulations issued in 1988. The regulations state that an employer has to make a 'suitable and sufficient assessment of the risks created by work liable to expose any employee to any substance hazardous to health'.

Less prescriptive requirements for risk assessment were contained in the Control of Lead at Work Regulations 1980 and the Control of Asbestos at Work Regulations 1987.

Is risk assessment as complicated as everyone seems to make out?

No. Risk assessment is a common sense approach to identifying the hazards and risks associated with a work task or activity.

Some knowledge of the task or job process is needed, as is some understanding of what could go wrong, what harm could be caused and to whom, and how likely it is to happen.

Risk assessments need to be carried out by 'competent people' and fully trained, degree-level educated experts are not always the best people for the job.

Some aspects of risk assessment are complicated, but these are usually where the work environment or processes undertaken are complex and potentially very dangerous (e.g. oil rigs, railways and chemical processing plants).

The Health and Safety Executive (HSE) states on its website that 'A risk assessment is not about creating huge amounts of paperwork, but rather about identifying sensible measures to control the risks in your workplace.'

What is meant by the term 'suitable and sufficient'?

Exactly what the words mean – suitable for the level and complexity of the job, and sufficient so that as many of the known hazards and risks as possible are identified.

A risk assessment needs to reflect the key elements of the job or activity and must consider all known consequences.

People are not expected to be able to 'see round corners', and sometimes the level of knowledge of a particular hazard or its potential has not yet been developed sufficiently (e.g. no one really knows at present whether the use of mobile technology equipment is a health hazard but it may be that future research will conclude that its use is a considerable health hazard). Employers are not expected to implement controls for hazards that are not fully understood.

'Suitable and sufficient' is based on the knowledge that exists at the time of undertaking the risk assessment.

What does 'so far as is reasonably practicable' mean?

It means balancing the level of risk against the measures needed to control the real risk in terms of money, time or trouble. However, you do not need to take action if it would be grossly disproportionate to the level of risk.

What does the term 'hazard' mean?

Hazard is the 'potential to cause harm'.

Virtually anything has the potential to cause harm if used inappropriately or in an unsuitable environment. Life is full of hazards – all manner of things *could* cause us harm (e.g. using electricity, crossing the road, driving the car, walking or playing sport).

Many work activities have the potential to cause harm to employees and others.

However, just because there is a hazard there may not be any harm caused because certain activities are undertaken to reduce the consequences of the hazard.

What is meant by the term 'risk'?

Risk is the likelihood that the hazard will be realised or the harm identified from the hazard will come to fruition.

Risk is what might happen – the consequences of the harm or hazard.

Risk can affect one person, several or hundreds, and the number of people that may potentially be affected helps to determine the seriousness of the risk.

What is meant by the term 'risk control'?

Risks are generally unacceptable – why accept the potential to be injured or harmed if you do not have to. Risk control is about managing the risks so that they become acceptable.

Risk control is the outcome of a risk management process.

What is meant by the term 'risk reduction'?

Risk reduction involves measures designed either to prevent hazards from creating risks or to lessen the distribution, intensity or severity of hazards.

The risk reduction assessment is a management tool that is used to reduce the likelihood and consequences of risks throughout an organisation.

Records, statistics, insurance losses, etc. can be reviewed to see what types of hazards and risks are present and the costs to the business. Consequent to the review an objective is set to reduce the risks of particular activities, processes, etc. so that the losses incurred will be reduced.

Risk reduction reduces the likelihood of things going wrong and reduces the consequences and costs arising from those risks.

What is risk avoidance?

Usually risk avoidance is a conscious decision on the part of the employer (or other) to avoid any risks associated with a job or activity.

Why accept the potential for harm if you do not have to? A risk avoidance strategy identifies hazards and risks and states that those risks are unacceptable and will not be tolerated. Some alternative way of completing the job or activity without incurring risk will be found.

Working at height is a hazard, and there is a risk of falling from the height, dropping objects or tools onto people below, etc. An employer may adopt a risk avoidance strategy by ensuring that all work at height is from a level platform with no risk of falling (e.g. windows are cleaned from the inside because they have been designed with 360° central pivots).

What is meant by the term 'risk profiling'?

The risk profile of an organisation informs all aspects of the approach to leading and managing its health and safety risks.

Every organisation will have its own risk profile. This is the starting point for determining the greatest health and safety issues for the organisation. In some businesses the risks will be tangible and immediate safety hazards, whereas in other organisations the risks may be health-related and it may be a long time before the illness becomes apparent.

A risk profile examines:

- the nature and level of the threats faced by an organisation
- the likelihood of adverse effects occurring
- the level of disruption and costs associated with each type of risk
- the effectiveness of controls put in place to manage those risks.

The outcome of risk profiling will be that the right risks have been identified and prioritised for action, and minor risks will not have been given too much priority. Risk profiling also informs decisions about what risk control measures are needed. Effort should be given to managing the most important risks (i.e. those that have the potential to create the most harm).

What regulations in respect of health and safety require risk assessments?

The concept of risk assessment has developed over the years and has become an important part of UK health and safety legislation.

New health and safety laws are generally 'self-regulatory' as opposed to being 'pre-scriptive'. This means that the regulations no longer tell employers what they must do and how to do it but set a framework for achieving 'goals' in respect of health and safety so that employers can arrive at a solution that best suits their businesses. The risk assessment approach enables this by, for instance, stating that employees will not be exposed to risk but allowing the employer to determine what the risk is and how to reduce it, etc.

The following regulations specifically require risk assessments:

- Control of Noise at Work Regulations 2005
- Manual Handling Operations Regulations 1992
- Health and Safety (First-Aid) Regulations 1981
- Health and Safety (Display Screen Equipment) Regulations 1992
- Personal Protective Equipment at Work Regulations 1992
- Regulatory Reform (Fire Safety) Order 2005
- Control of Substances Hazardous to Health Regulations 2002
- Control of Asbestos Regulations 2012
- Control of Lead at Work Regulations 2002
- Dangerous Substances and Explosive Atmospheres Regulations 2002
- Ionising Radiations Regulations 1999
- Control of Major Accident Hazards Regulations 2015.

In addition, the Management of Health and Safety at Work Regulations 1999 set out the framework and expectation for all risk assessments. If risk assessments are not called for under specific regulations the requirement to complete them in any event is contained in the Management Regulations.

What are the consequences for failing to carry out risk assessments?

The ultimate consequence for failing to carry out a risk assessment is that an employee or other person could be killed by the job or activity you have asked them to carry out as their employer.

Accidents and injuries to employees or the general public caused by work activities can lead to compensation claims through the civil courts or insurance companies.

Prosecutions for failing to have suitable and sufficient risk assessments are common, and, depending on the circumstances of the case, the case can be heard in the magistrates' court or the Crown Court.

Fines imposed by the magistrates' court can be unlimited for each missing or inadequate risk assessment, and if a serious breach of the law occurs because of a missing or inadequate risk assessment the case could be heard in the Crown Court, where fines are unlimited (e.g. if the lack of a risk assessment contributes to a fatality).

What is the difference between a risk assessment and a method statement?

A risk assessment is simply a careful examination of what, in your work, could cause harm to people, so that you can weigh up whether you have enough precautions or whether you should undertake more.

As an employer or self-employed person, you must do a risk assessment, but you only need to record it if you employ five or more people.

A safety method statement is not required by law. It describes in a logical sequence exactly how a job is to be carried out in a safe manner and without risks to health. It includes all the risks identified in the risk assessment and the measures needed to control those risks. This allows the job to be properly planned and resourced. The method statement is the safe system of work, and is a way to control risks identified in the risk assessment.

Safety method statements are most often found in the construction sector. They are particularly helpful for:

- higher risk, complex or unusual work (e.g. steel and formwork erection, demolition or the use of hazardous substances)
- providing information to employees about how the work should be done and the precautions to be taken
- providing the principal contractor with information to develop the health and safety plan for the construction phase of a project.

Whether or not safety method statements are used it is essential to make sure that risks are controlled.

What is the legal liability of individual board members for health and safety failures?

If a health and safety offence is committed with the consent or connivance of, or is attributable to any neglect on the part of, any director, manager, secretary or other similar officer of the organisation, then that person (as well as the organisation) can be prosecuted under Section 37 of the Health and Safety at Work etc. Act 1974.

Case law has confirmed that directors cannot avoid a charge of neglect under Section 37 by arranging their organisation's business so as to leave them ignorant of circumstances that would trigger their obligation to address health and safety breaches.

Those found guilty are liable for fines and imprisonment. In addition, the Company Directors Disqualification Act 1986, Section 2(1), empowers the court to disqualify an individual convicted of an offence in connection with the management of a company. This includes health and safety offences. This power is exercised at the discretion of the court; it requires no additional investigation or evidence.

Individual directors are also potentially liable for other related offences, such as the common law offence of gross negligence manslaughter. Under the common law, gross negligence manslaughter is proved when individual officers of a company (directors or business owners) by their own grossly negligent behaviour cause death. This offence is punishable by an unlimited fine and a maximum of life imprisonment.

Do I need to employ a consultant to carry out my risk assessments?

No, not unless the work processes are complex, or you feel unable to carry out the task yourself, or you don't have anyone at your workplace capable of carrying them out.

Risk assessments are not complicated or difficult to do but they do need a certain amount of knowledge about hazard and risk, legal requirements for health and safety, and knowledge of the work activity being carried out.

Often the best people to help you complete risk assessments are your employees – remember you are looking out for anything that happens in your workplace which could cause injury or ill-health.

Consultants can provide a wealth of knowledge and experience, and may be able to make the whole process simple and painless. On the other hand, they could overcomplicate things!

Are generic risk assessments acceptable?

The purpose of a risk assessment is to identify hazards and risks associated with specific tasks carried out in specific work locations.

If all the tasks and all the locations are the same, then one risk assessment should suffice for all activities and a generic assessment will be sufficient. But, if either the task itself or the location differs in any way from place to place, then a 'site-specific' risk assessment will be needed.

Generally, generic risk assessments are seen by enforcing authorities as mere 'tick box' exercises that do not add much to the management of risk.

Remember, your risk assessment is to help you, as the employer, to prevent the likelihood of injury or ill-health to your employees. So, you really need to review exactly what your employees do and how they do it – not what you think they do (as often this is not the same – bad habits, etc.).

Who can carry out risk assessments?

Whoever carries out a risk assessment must be 'competent' to do the job. Competency is not defined in the legislation, but such persons must have suitable training and experience or knowledge in order to identify hazards and risks associated with the job tasks. Managers, employees, consultants, individuals or teams can undertake risk assessments.

Complex work processes and tasks may require the assistance of external specialist consultants in undertaking the risk assessment. But the *employer* is always responsible for ensuring that a suitable and sufficient risk assessment is completed.

Risk assessments could be divided into different work or departmental areas so that those who know what happens in their department use their expertise to complete their own assessments.

The HSE states on its website that many organisations, where they are confident they understand what's involved, can do the assessment themselves. You do not have to be a health and safety expert.

When should risk assessments be reviewed?

The Management of Health and Safety at Work Regulations 1999 require a risk assessment to be reviewed if:

■ there is reason to suggest that it is no longer relevant or valid
■ there has been significant change in the matters to which it relates.

If changes to the risk assessment are required, the employer has a duty to make the changes and reissue the risk assessment.

Employers are not expected to anticipate risks that are not foreseeable.

However, if events occur that alter the information available or the perception of risk, the employer will be expected to respond to the new information and assess the hazards and risks in light of the increased knowledge.

Accidents and near misses should be investigated, as these incidents will indicate whether more knowledge is available on the hazard or risk associated with the job. The risk assessment may need to be reviewed because:

■ something previously unforeseen has occurred
■ the risk of something happening or the consequences of the event may be greater than expected or anticipated
■ precautions prove less effective than anticipated.

New equipment, a new working environment, new materials, different systems of work, etc. will all require existing risk assessments to be reviewed.

Case studies

Director should have known

The managing director of a manufacturing company with around 100 workers was sentenced to 12 months' imprisonment for manslaughter following the death of an employee who became caught in unguarded machinery. The investigation revealed that, had the company adequately maintained guarding around a conveyor, the death would have been avoided.

The judge made clear that whether the managing director was aware of the situation or not, he should have known, as this was a long-standing problem. An area manager also received a custodial sentence. The company received a substantial fine and had to pay the prosecution costs.

Failure to train leads to unsafe system of work

A company and its directors were fined a total of £245 000 and ordered to pay costs of £75 500 at Crown Court in relation to the removal of asbestos. The company employed ten, mostly young, temporary workers, who were not trained or equipped to remove the asbestos safely, nor warned of the risk involved. The directors were also disqualified from holding any company directorship for 2 years and 1 year, respectively.

Lack of risk assessments led to deaths

Two farm workers were killed while they were using a tractor-mounted hydraulically driven machine used for winding up long lengths of rope. They became entangled between the rope and a rotating shaft. The machine was being used to wind up long lengths of rope used to secure polythene onto the metal hoops of polytunnels, which protect the fruit grown on the farm. The standard practice was for one employee to stand about 6 m behind the winding machine and feed the rope, while the other stayed in the stationary tractor cab operating the controls. It is not known why the tractor driver left the cab without first turning off the machine. The bodies of both workers were found some hours after the accident.

The investigation concluded that the deceased men were using a tractor-mounted fleece-winder machine, which was not suitable for rope winding as, due to its design, it had a significant rundown time after it was switched off and it did not have an automatic cut-off and braking system in the event of entanglement.

The employees had not been made aware of the dangers posed by the task, and had not been adequately trained. This resulted in a tragic and preventable accident, which resulted in the loss of two young lives.

Both the partnership which ran the farm, owned the machine and employed one of the men, and the specialist company that was contracted to dismantle the tunnels were prosecuted.

The farming partnership was prosecuted under: (i) Section 2(1) of the Health and Safety at Work etc. Act 1974, for failing to ensure the safety at work of employees by the provision of safe plant and safe systems of work, and (ii) Regulation 3(1) of the Management of Health and Safety at Work Regulations 1999, for failing to undertake a suitable and sufficient assessment of the risk to the health and safety of employees. The company contracted to build and dismantle the polytunnels was prosecuted under Section 3 of the Health and Safety at Work etc. Act 1974 for failing to produce a risk assessment.

Both companies pleaded guilty. The farming partnership was fined a total of £60 000 plus prosecution costs of £45 548, and the contractor was ordered to pay £20 000 plus prosecution costs of £15 516.

Risk Assessments: Questions and Answers
ISBN 978-0-7277-6076-0

ICE Publishing: All rights reserved
http://dx.doi.org/10.1680/raqa.60760.013

Chapter 2
Legal framework

What is the main piece of legislation which sets the framework for health and safety at work?

The Health and Safety at Work etc. Act 1974 is the main piece of legislation that sets out the broad principles of health and safety responsibilities for employers, the self-employed, employees and other persons.

The Health and Safety at Work etc. Act 1974 is known as an 'enabling act', as it allows subsidiary regulations to be made under its general enabling powers. It sets out 'goal objectives', and was one of the first pieces of legislation to introduce an element of self-regulation.

The Act place responsibilities on employers to:

- safeguard the health, safety and welfare of employees
- provide a safe place of work
- provide safe equipment
- provide safe means of egress and access
- provide training for employees
- provide information and instruction
- provide a written safety policy
- provide safe systems of work.

The Act places responsibilities on employees to:

- cooperate with their employer in respect of health and safety matters
- wear protective equipment or clothing if required
- safeguard their own and others' health and safety
- not recklessly or intentionally interfere with or misuse anything provided in the interests of health and safety or welfare
- not to tamper with safety equipment provided by their employers for the safety of themselves or others.

The Act places responsibilities on 'persons in control of premises' to:

- ensure safe means of access and egress
- ensure that persons using premises who are not their employees are reasonably protected in respect of health and safety.

The Act also requires employers to conduct their undertaking in such a way that persons who are not their employees are not adversely affected by it in respect of health and safety.

Finally, the Act requires manufacturers and suppliers and others who design, import, supply, erect or install any article, plant machinery, equipment or appliance for use at work, or who manufacture, supply or import a substance for use at work, to ensure that health and safety matters are considered in respect of their product or substance.

The term 'reasonably practicable' is used in the Health and Safety at Work etc. Act 1974 and numerous regulations. What does it mean?

The term 'reasonably practicable' is not defined in any of the legislation in which it occurs. Only the courts can make an authoritative judgement on what is reasonably practicable.

Case law has been built up over the years and practical experience gained in interpreting the law. A common understanding of 'reasonably practicable' is:

the risk to be weighed against the costs necessary to avert it, including time and trouble as well as financial costs.

If, compared with the costs involved of removing or reducing it, the risk is small (i.e. consequences are minor or infrequent), then the precautions need not be taken.

Any establishment of what is reasonably practicable should be taken before any incident occurs.

The burden of proof in respect of what is reasonably practicable in the circumstances rests with the employer or other duty holder. Employers or other duty holders would generally need to prove why something is *not* reasonably practicable at a particular point in time.

An ability to meet the costs involved in mitigating the hazard and risk is *not* a factor that the employer can take into consideration when determining 'reasonably practicable'. Costs can only be considered in relation to whether it is reasonable to spend the money given the risks identified.

It is only possible to determine 'reasonably practicable' when a full, comprehensive risk assessment has been completed.

How can civil action under civil law be actioned by employees in respect of health and safety at work?

Civil action can be initiated by an employee who has suffered injury or damage to health caused by his or her work.

The employer may be in breach of the 'duty of care' that he owes to the employee. The employer may have been negligent in common law – the body of law that has been determined by case law and that has evolved rather than been set down by parliament.

Employers could face civil claims because they have breached the law and irrespective of whether the employee has been negligent in ensuring his or her own or others' safety.

However, in order to ensure that responsible employers are not unfairly penalised for injuries at work, Section 69 of the Enterprise and Regulatory Reform Act 2013 came into force in October 2013, resulting in a significant change to employer liability personal injury claims. The effect of Section 69 is that most workers seeking compensation for injuries suffered as a result of accidents at work on or after 1 October 2013 will no longer be able to rely solely on a breach of health and safety regulations to establish liability. Instead, they will only be able to seek compensation where it can be shown that the employer was actually at fault or negligent.

How quickly do employees need to bring claims to the courts in respect of civil claims?

Civil actions must commence within 3 years from the time of knowledge of the cause of action. This will usually be the date on which the employee knew or should have known that there was a significant injury and that it was caused by the employer's negligence.

Therefore, an employer would be wise to keep all records of training, risk assessments, checks, maintenance schedules, etc. for a minimum of 3 years, as these may be needed for any defence to a claim, etc.

A civil claim will succeed if the plaintiff – the person bringing the case – can prove breach of statutory duty or the duty of care beyond 'the balance of probabilities'.

The employer may mount a number of defences to the claim, the most common of which are

■ contributory negligence – i.e. the injured employee was careless or reckless (e.g. ignored clear safety rules and procedures)

- the injuries were not reasonably foreseeable – i.e. the injuries were beyond normal expectation or control – the employer did not have the knowledge to foresee the risks, neither did science or experts
- voluntary assumption of risk – i.e. the employee consents to take risks as part of the job – but the employer cannot rely on this defence to excuse him of fulfilling his duties under legislation (no one can contract out of their statutory duties).

What are the consequences for the employer if an employee wins a civil action?

Employers will invariably be required to pay damages or compensation. Compensation claims can run into thousands of pounds in some cases. The employer's liability insurance will cover the cost of the claims, less any excess that the employer has opted to pay.

Damages are assessed on:

- loss of earnings
- damage to any clothing, property or personal effects
- pain and suffering
- future loss of earnings
- disfigurement
- inability to lead an expected, normal personal or social life because of the injury, among others
- medical and nursing expenses.

Not only will there be the financial pay-out but also associated bad publicity, which could lead to loss of reputation.

Who enforces health and safety legislation?

There are in the main two organisations with powers to enforce health and safety legislation:

- the Health and Safety Executive (HSE)
- local authorities.

The HSE enforces the law in the following types of work environment:

- industrial premises
- factories, manufacturing plant
- construction sites
- hospitals and nursing/medical homes
- local authority premises

- mines and quarries
- railways
- broadcasting, filming
- agricultural activities
- shipping
- airports
- universities, colleges and schools.

Local authorities, usually through their environmental health departments are responsible for:

- retail sale of goods
- warehousing of goods
- exhibitions
- office activities
- catering services
- caravan and camping sites
- consumer services provided in a shop
- baths, saunas and body treatments
- zoos and animal sanctuaries
- churches and religious buildings
- childcare businesses
- residential care (although not nursing care).

The HSE and local authorities have the same powers but the way they use them may differ. Local authority inspectors may visit premises more frequently than HSE inspectors but the HSE may be tougher on taking formal action, including the issuing of fee for intervention penalty notices.

What powers do enforcement authorities have to enforce health and safety legislation?

Inspectors can take action when they encounter a contravention of health and safety legislation and when they discover a situation where there is imminent risk of serious personal injury.

An inspector can also instigate legal proceedings, although in reality the decision to proceed to court is often taken by the enforcement agency's in-house legal team.

An inspector may serve an *improvement notice* if he or she is of the opinion that a person:

- is contravening one or more of the relevant statutory provisions or

- has contravened one or more of those provisions in circumstances where the contravention is likely to occur again or continue.

The inspector must be able to identify that one or more legal requirements under acts or regulations is being contravened (e.g. failure to complete a risk assessment, or operating an unsafe system of work).

An improvement notice must:

- state what 'statutory provisions' are, or have, been contravened
- state in what way the legislation has been contravened
- specify the provisions that must be taken to remedy the provision
- specify a time within which the person is required to remedy the contraventions.

The time to remedy contraventions highlighted in an improvement notice must be at least 21 days. This is because there is an appeal procedure to the service of an improvement notice and the appeal must be brought within 21 days.

An improvement notice served on a company as an employer must be served on the registered office and is usually served on the company secretary.

If an inspector believes that health and safety issues are so appallingly managed in workplaces or premises, etc. and that there is 'a risk of serious personal injury', then a *prohibition notice* can be served. The inspector must be able to show or prove a risk to health and safety.

Prohibition notices can be served in *anticipation* of danger, and the inspector does not have to identify specific health and safety legislation that is being, or has been, contravened.

A prohibition notice must:

- state the inspector's opinion that there is a risk of serious personal injury
- specify the matters that create the risk
- state whether statutory provisions are being, or have been or will be contravened, and, if so, which ones
- state that the activities described in the notice cannot be carried on by the person on whom the notice is served, unless the provisions listed in the notice have been remedied.

A prohibition notice takes effect immediately where stated, or can be 'deferred' to a specified time.

Risks to health and safety do not need to be imminent, but usually there must be a hazard that is likely to cause imminent risk of injury, otherwise the inspector could serve an improvement notice.

What are the consequences for failing to comply with any of the notices served by inspectors?

Failure to comply with either an improvement notice or a prohibition notice is an offence under the Health and Safety at Work etc. Act 1974.

Legal proceedings are issued against the person, employer or company on whom the notice was served, and the matter will be heard in the magistrates' court in the first instance. However, serious contraventions can be passed to the Crown Court, where powers of remedy are greater.

Failure to comply with an improvement notice carries an unlimited fine in both the magistrates' court and the Crown Court. Custodial sentences are also possible.

Failure to comply with a prohibition notice carries an unlimited fine in the magistrates' court and in the Crown Court. Imprisonment of persons who contravene a prohibition notice is also an available option in both the magistrates' court and the Crown Court.

Where employers or others fail to comply with a prohibition notice and continue, for example, to use defective machinery, and where, as a consequence of using that defective machinery a serious accident occurs, there will invariably be a prosecution, and the courts are likely to take a serious view and hand down prison sentences.

Can an inspector serve a statutory notice and prosecute at the same time?

Yes. If an inspector serves an improvement notice or prohibition notice he or she may decide that the contravention is so blatant or so serious that immediate prosecution is warranted. The notices ensure that unsafe conditions are remedied during the prosecution process, as legal cases can take several months to come before the courts.

Employers and others do not have to be given time to remedy defects once they have been identified, as many duties on employers are 'absolute' (i.e. the employer 'must' do something).

Inspectors will often prosecute without initiating formal action where they believe they have given the employer ample opportunity to put right defects (e.g. they may have given advice during a routine inspection or issued informal letters).

When accidents are being investigated, contraventions of health and safety legislation will be viewed seriously, and often prosecutions will be taken.

What is the appeal process against the service of improvement and prohibition notices?

Any person on whom a notice is served can appeal against its service on the grounds that:

- the inspector wrongly interpreted the law
- the inspector exceeded his powers
- the proposed solution to remedy the default is not practicable
- the breach of law is so insignificant that the notice should be withdrawn.

Lodging an appeal must be made within 21 days of the service of the notice.

An appeal is to an employment tribunal.

An improvement notice is suspended pending the appeal process. This means that the employer or person served with the notice will be able to continue doing what they have been doing without changing practices or procedures.

Costs for bringing an appeal may be awarded by the tribunal – either in favour of the enforcement authority if the appeal is dismissed, or for the employer if the appeal is successful.

A prohibition notice may continue in force pending an appeal unless the employer or person in receipt of the notice requests the employment tribunal to suspend the notice pending the appeal.

All statutory notices served by enforcement authorities should contain information on how to lodge an appeal.

On hearing the appeal, an employment tribunal can:

- dismiss the appeal, upholding the notices as served
- withdraw the notices, thereby upholding the appeal
- vary the notices in respect of the time given to complete works (i.e. the remedies listed in any schedule)
- impose new remedies not contained in the notice if these will provide the solutions necessary to comply with the law.

If remedial works cannot be completed in the time given in the improvement notice, can the time be extended?

Yes. It is allowable for the inspector who served the notice to extend the time limits given if works cannot be completed in the time specified, and a request is submitted to the enforcing authority.

It would be prudent to explain why compliance cannot be achieved, what steps have been taken in the interim and when compliance is expected.

What are the powers of inspectors under the Health and Safety at Work etc. Act 1974?

Enforcing authorities and their individual inspectors have wide-ranging powers under the Act, and in addition to the service of notices inspectors can:

- enter and search premises
- seize or impound articles, substances or equipment
- instruct that premises and anything in them remain undisturbed for as long as necessary while an investigation is conducted
- take measurements, photographs, recordings
- detain items for testing or analysis
- interview people
- take samples of anything for analysis, including air samples
- require to see and copy, if necessary, any documents, records, etc. relevant to the investigation or inspection
- require that facilities are made available to them while carrying out their investigations.

If inspectors believe they will meet resistance they may be accompanied by a police officer. It is an offence to obstruct inspectors while they are carrying out their duties.

What are the fines for offences against health and safety legislation?

Health and safety offences are usually 'triable either way', which means that they can be heard in a magistrates' court or the Crown Court. There are some relatively minor offences that can only be heard in a magistrates' courts and some serious offences that can only be heard in the Crown Court. The Health and Safety (Offences) Act 2008 increased various health and safety fines and extended the number of offences that can be punished by custodial sentences.

The sentencing powers of the two courts are different, with the Crown Court operating with a jury. Higher fines and imprisonment can be imposed by the Crown Court, which historically has dealt with the more serious cases involving death and major injury or widespread potential harm.

Offences only triable in a magistrates' court are:

- contravening an investigation ordered by the Health and Safety Commission
- failing to answer questions under Section 20 of the Health and Safety at Work etc. Act 1974
- obstructing an inspector in his or her duties
- preventing another person from answering questions or cooperating with an inspector
- impersonating an inspector.

Offences that are 'triable either way' are:

- failure to comply with any or all of Sections 2–7 of the Health and Safety at Work etc. Act 1974
- contravening Section 8 of the 1974 Act – intentionally or recklessly interfering with anything provided for safety
- levying payment on employees or others for safety equipment, etc. contrary to Section 9 of the 1974 Act
- contravening any of the health and safety regulations made under the 1974 Act or other enabling legislation
- contravening the powers of inspectors in relation to the seizure of articles, etc.
- contravening the provisions and requirements of improvement and prohibition notices
- making false declarations, keeping false records, etc. in relation to health and safety matters.

If a case is to be brought before a magistrates' court (a summary offence) it must be brought within 6 months from the date the complaint is laid (i.e. lodged at the magistrates' court and summonses issued).

Legislation came into effect on 12 March 2015 providing for all maximum fines in magistrates' courts of £5000 or more to become unlimited in England and Wales. The legislation is contained in Section 85 of the Legal Aid, Sentencing and Punishment of Offenders Act 2012.

Previously, fines payable on conviction in a magistrates' court were capped, often at a statutory maximum of £5000 on the magistrates' court 'standard scale', or a higher amount, if so provided for in the applicable legislation. For example, health and safety offences under the Health and Safety at Work etc. Act 1974 were capped at £20 000. Fines will be imposed for:

- Breaches of Sections 2–6 of the Health and Safety at Work etc. Act 1974:
  - magistrates' court – an unlimited fine for each offence, or imprisonment for a maximum of 6 months, or both
  - Crown Court – an unlimited fine for each offence, or imprisonment for up to 2 years, or both.
- Breaches of improvement and prohibition notices:
  - magistrates' court – an unlimited fine for each offence, or imprisonment for up to 6 months, or both
  - Crown Court – unlimited fine, or imprisonment for up to 2 years, or both.
- Breaches of health and safety regulations and other sections of the 1974 Act:
  - magistrates' court – an unlimited fine per offence and/or imprisonment for up to 6 months
  - Crown Court – unlimited fines per offence, or imprisonment for up to 2 years, or both.

Guidelines were issued in late 2015 by the Sentencing Council for England and Wales recommending that fines for breaches of health and safety law be levied in accordance with a company's ability to pay (i.e. based on the company's turnover and available assets). There is likely to be an increase in the level of unlimited fines in both the magistrates' court and the Crown Court; that fines will run in to tens and thousands of pounds so as to act as real deterrents to businesses in respect of breaking the law.

Prosecutions for health and safety offences 2014–2015

According to its annual report, in 2014–2015 the HSE:

- prosecuted 650 cases, with at least one conviction achieved in 606 cases, a conviction rate of 93%
- prosecuted 1058 offences, resulting in 905 convictions, a conviction rate of 86%,

and

- prosecutions led to fines totalling £16.7 million, an average penalty of £18 944 per offence.

In the 2014/2015 year local authorities:

- prosecuted 78 cases, with a conviction secured in 76 cases, a conviction rate of 97%
- prosecuted 153 offences, resulting in 142 convictions, a conviction rate of 96%
- prosecutions led to fines totalling £2.8 million, an average penalty of £19 962 per offence.

What is a Section 20 interview under the Health and Safety at Work etc. Act 1974?

Under Section 20 inspectors can require anyone to answer questions as they think fit in relation to any actual or potential breach of the legislation, accident or incident investigation, etc.

A Section 20 interview is, or should be, a series of questions asked by the inspector – not a witness statement where the interviewee describes what happens.

The answers to Section 20 questions must be recorded and the interviewee must sign a declaration that they are true.

However, evidence given in a Section 20 interview is inadmissible in any proceedings subsequently taken against the person giving the interview or statement, or his or her spouse.

If an inspector is contemplating bringing criminal proceedings, the person will usually be interviewed under the Police and Criminal Evidence Act 1984 (PACE), as the information gathered during these interviews is admissible in court. The code of practice on PACE interviews is strict (e.g. a caution must be given and a failure to follow the procedures could result in acquittal on the grounds of a technicality).

The HSE issues 'guidance' and codes of practice. What are these and is a criminal offence committed if they are not followed?

The HSE endeavours to provide employers and others with as much information as possible on how to comply with legislation.

Guidance documents are issued on a variety of specific industries or particular processes with the purpose of:

- interpreting the law (i.e. helping people to understand what the law says)
- assisting with complying with the law
- giving technical advice.

Following guidance is not compulsory on employers, and they are free to take other actions to eliminate or reduce hazards and risks.

However, if employers do follow the guidance as laid down in the HSE documents, they will generally be doing enough to comply with the law.

Approved codes of practice (known as ACOPs) are the other common documents issued by the HSE. These set out good practice and give advice on how to comply with the law.

A code of practice will usually illustrate the steps that need to be taken to be able to show that 'suitable and sufficient' steps have been taken in respect of managing health and safety risks.

ACOPs have a special legal status. If employers are prosecuted for a breach of health and safety law and it is proved that they have *not* followed the relevant provisions of the ACOP a court can find them at fault unless they can show that they have complied with the law in some other way.

If an employer can show that it followed the provisions of an ACOP it will be unlikely to be prosecuted for an offence. Equally, if an employer follows the ACOP and the enforcing authority serves an improvement or prohibition notice, the employer would have grounds to appeal the notice.

What is the HSE's fee for intervention scheme?
The HSE introduced the fees for intervention (FFI) scheme on 1 October 2012.

Under the FFI scheme, employers contravening health and safety law must pay the HSE's enforcement costs at a rate of £124 per hour until the breach has been rectified. Invoices are rendered every 2 months.

In October 2010, the Department for Work and Pensions announced that, following a comprehensive spending review, it was cutting the HSE's grant by 35% over 4 years from April 2011, this amount being roughly equivalent to £80 million. To recoup some of this loss, the government introduced the FFI scheme to recover the costs from businesses failing to comply with health and safety regulation.

The scheme applies to all businesses and organisations inspected by the HSE. Inspections by other regulators, such as local authority environmental health officers, are not affected.

The FFI does not apply to businesses already paying fees to the HSE for its work through other arrangements, such as control of major accident hazards (COMAH) charges.

The intention is that the HSE recovers the cost of its regulatory work from duty holders found to be in 'material breach' of health and safety law. This will be either from a proactive or a reactive visit where the inspector judges that there has been a material breach serious enough for him or her to notify the duty holder in writing of that contravention.

What is a 'material breach'?

A material breach is defined in HSE guidance as a situation where an inspector is 'of the opinion that there is or has been a contravention of health and safety law that requires him to issue notice in writing to that effect'; therefore, it includes a letter, an improvement or prohibition notice or a prosecution. Before deciding to notify the duty holder, an inspector must apply the principles of the HSE's Enforcement Policy Statement and its Enforcement Management Model and have regard to the guidance issued (all of which can be accessed via the HSE website) to ensure that the decision taken about the level of any enforcement action is proportionate in all the circumstances.

Any notification must make clear which contraventions are considered material breaches.

Inspectors have invoiced for the time they have spent identifying the breach, advising on putting it right, investigating and taking enforcement action. All the time spent on carrying out visits, being on the site during which the material breach was identified, the writing of letters, notices and reports, taking statements and getting specialist reports for complex issues is included.

Chargeable time runs from the start of the visit when the material breach is identified until it is corrected. Payment is due within 30 days. Non-payment is a civil debt, not a criminal penalty.

There is a dispute procedure in place. If a dispute is raised the enforcement action taken will be reviewed by a senior manager in the HSE. If you are not satisfied with the response, then there is the possibility for a further appeal to a panel comprising two HSE staff and an independent representative.

If this appeal is unsuccessful, you will have to pay for the HSE's time spent in dealing with the dispute at the FFI hourly rate. If the dispute is upheld, the HSE will refund any invoices or part invoices that have been paid.

What are the consequences for failing to comply with any of the notices served by inspectors?

Failure to comply with either an improvement notice or a prohibition notice is an offence under the Health and Safety at Work etc. Act 1974.

Legal proceedings are issued against the person, employer or company on which the notice was served and the matter will be heard in a magistrates' court in the first instance. However, serious contraventions can be passed to the Crown Court, where powers of remedy are greater.

Failure to comply with an improvement notice carries an unlimited fine in the magistrates' court, or an unlimited fine in the Crown Court, plus possible imprisonment.

Failure to comply with a prohibition notice carries an unlimited fine in the magistrates' court, or an unlimited fine in the Crown Court. Imprisonment of persons who contravene a prohibition notice is also an available option for the courts - both the magistrates' court and Crown Court.

Where employers or others fail to comply with a prohibition notice and continue, for example, to use defective machinery, and where, as a consequence of using that defective machinery, a serious accident occurs, there will invariably be a prosecution, and the court is likely to take a serious view and hand down prison sentences.

What do I need to know about the Corporate Manslaughter and Corporate Homicide Act 2007?

Corporate manslaughter is a crime that can be committed by a company in relation to work-related deaths.

The Act enables a company to be prosecuted for negligent conduct that leads to a person's death.

Every offence has a legal test:

- The accused had a legal duty to care for the deceased.
- The accused breached the duty of care.
- The breach of duty caused the death and was so severe a breach of duty to be a crime (i.e. gross negligence).

If the breach of duty falls considerably below what would be expected of the organisation it will be classed as gross negligence.

Consideration is given to:

- how serious the failure was
- how real the risk of death was
- the attitudes, policies and accepted practices of the organisation which led to the failure
- any health and safety guidance relating to the breach.

In broad terms, the prosecution has to be able to prove that there were management failings in respect of health and safety standards.

Penalties are unlimited and must reflect the severity of the case and the affordability to the accused. In addition, publicity orders can be imposed requiring the company to publicise the case and the actions it has taken to remedy its gross negligence.

What information might be relevant for inclusion in the company's annual report?

You should include appropriate health and safety information in your published reports on your activities and performance in respect of health and safety. This demonstrates to your stakeholders your commitment to effective health and safety risk management principles.

Good corporate governance requires all companies to review their risk management policies, including health and safety. The Turnbull Report, *Internal Control: Guidance for Directors on the Combined Code* (1999), summarised the approach to be taken on this subject.

As a minimum, the company's annual report should contain the following in respect of health and safety information:

- the broad context of the policy on health and safety, including significant risks faced by staff and the arrangements for consulting employees and involving safety representatives
- the company's health and safety goals, including specific and measurable targets for improving health and safety within the organisation.

Some specific and measurable targets may include:

- reducing working days lost due to injury or ill-health by x% per 100 000 worker hours
- reducing the incidence rate of work-related ill-health by x% per number of employees
- reducing the major injury or fatality rate by x% by the year 2010
- the number of reportable accidents, ill-health and dangerous occurrences notified to the authorities under the Reporting of Injuries, Diseases and Dangerous Occurrences Regulations 1995
- details of any major accidents and fatalities, and the improvements implemented as a result of the ensuing investigation.
- the total number of staff days lost due to accidents or ill-health caused by work activities
- the number of health and safety enforcement notices served on the company within the reporting period

- the number and nature of convictions for health and safety offences
- the continuous improvement initiatives undertaken by the organisation
- the outcome of health and safety audits, internal and external monitoring processes, etc.
- the investment undertaken in respect of training and information initiatives for employees in respect of health and safety subjects, etc.

Increasingly, stakeholders in a business, including investors, want to have a clear understanding of the integrity, ethics and values of the companies they are dealing with. The new approach of 'corporate social responsibility' requires companies to be transparent with regard to their performance in all areas of social, cultural and ethical dealings.

As an employer, am I responsible for the actions of my employees if they cause an accident to happen that results in injury to either themselves or others?

Yes, employers are responsible if persons are injured by the wrongful acts of their employees, if such acts are committed in the course of their employment.

This is known as 'vicarious liability'. *Vicarious liability* refers to a situation where someone is held responsible for the actions or omissions of another person. In a workplace context, an employer can be *liable* for the acts or omissions of its employees, provided it can be shown that these took place in the course of their employment.

There is no vicarious liability if the act was not committed in the course of employment (e.g. one employee assaulting another is not something the employer would be liable for, although a case ruling in the Court of Appeal in 2012 stated that, where an employee inflicts violence on another employee or third party, there can be vicarious liability of the employer for the employee's violent act where the incident is linked somehow to the employee's job).

What is the 'duty of care'?

Every member of society is under a duty of care for something or someone.

Duty of care really means to take reasonable care to avoid acts or omissions that it can be reasonably foreseen are likely to injure a neighbour or anyone else who ought reasonably to be kept in mind.

Employers own a duty of care to their employees.

Employers also owe a duty of care to contractors, visitors, members of the public, people on neighbouring properties, etc.

The duty of care owed by employers to employees includes:

- safe premises
- safe systems of work
- safe plant, equipment and tools
- safe fellow workers.

Other legislation also implies a duty of care on people, namely:

- Occupiers Liability Act 1957 and 1984
- Consumer Protection Act 1987.

What is a safety policy?

Under the Health and Safety at Work etc. Act 1974 employers must produce a written health and safety policy if they have five or more employees.

The policy must contain a written statement of the general policy on health and safety, the organisation of the policy and the arrangements for carrying it out.

Employees must be made aware of the safety policy and must be given information, instruction and training in its content, use, their responsibilities, etc.

A copy of the safety policy statement must either be given to all employees or be displayed in a prominent position in the workplace.

The safety policy must be reviewed regularly by the employer and kept up to date to reflect changes in practices, procedures, the law, etc.

What is meant by 'organisational arrangements'?

This section of the safety policy shows how the organisation will put its good intentions into practice, and outlines the responsibilities for health and safety for different levels of management within the company.

An organisational section will normally include:

- health and safety objectives
- responsibilities of
  - managing director or chief executive officer
  - operations director
  - safety director
  - senior management

  - departmental heads
  - maintenance
  - employees
- training arrangements
- monitoring and review processes
- appointment of competent persons
- consultation process for health and safety
- appointment of employee representatives
- procedures for conducting risk assessments
- emergency plans.

What is meant by 'arrangements' in a safety policy?

The safety policy must either contain details of what employees and others must do in order to ensure their safety at work, or contain references as to where information on safe practices can be found (e.g. in the department handbook or employee induction pack).

Usually, however, for ease of use and clarity, most employers will collate everything needed for the safety policy into one document.

The 'arrangements' section of the safety policy contains the details of *how* you expect your employees and others (e.g. contractors) to proceed safely with a task or job activity.

Subjects often covered in the arrangements are:

- accident and incident reporting and investigations
- first aid
- risk assessments
- fire risk assessments
- manual handling
- using equipment
- electricity and gas safety
- personal protective equipment
- emergency procedures
- fire safety procedures
- training
- monitoring and review procedures
- control of substances hazardous to health procedures
- occupational health
- maintenance and repair
- permit to work procedures

- stress in the workplace
- violence in the workplace
- operational procedures.

A safety policy needs to be 'suitable and sufficient', not perfect.

If an accident happens in the workplace the investigating officer (either an HSE investigator or an environmental health officer) will almost always want to see copies of the safety policy and risk assessments. Investigating officers will be looking to see if you have considered the hazard and risks of the job and implemented control measures. They will want to establish whether employees knew what to do safely, and the best place to review such information is in the safety policy.

Prosecutions have been taken for failing to have a written safety policy and also for having a totally inadequate one.

The safety policy should be thought of as the tool of communication between the employer and the workforce. It should be the reference guide on how employees are expected to perform their job tasks safely.

What are the five steps to successful health and safety management?

The five steps are:

Step 1: Set your policy
Step 2: Organise your staff
Step 3: Plan and set standards
Step 4: Measure your performance
Step 5: Audit and review.

Step 1: Set your policy

- Do you have a clear health and safety policy?
- Is it written down?
- Is it up to date?
- Does it specify who is responsible for what and who has overall safety responsibility?
- Does it give responsibilities to directors, and is there a clear commitment that the health and safety culture starts at the top?
- Does it specify arrangements for carrying out risk assessments, identifying hazards and implementation of control measures?
- Does it name competent persons?

- Does it state how health and safety matters will be communicated throughout the organisation?
- Does it have safety objectives?
- Has it had a beneficial effect on the business?
- Does it imply a proactive safety culture within the organisation?

Step 2: Organise your staff
- Have your involved your staff in your health and safety policy?
- Are you 'walking the talk'?
- Is there a health and safety culture?
- Have you adopted the 'four Cs':
  - competence
  - control
  - cooperation
  - communication?
- Are your staff and others competent to do their jobs safely?
- Are they properly trained and informed?
- Are there competent people around to help and guide them?
- Have you designated key people responsible for safety in each area of the business?
- Do employees know how they will be supervised in respect of health and safety?
- Do employees know who to report faults and hazards to and what will be done?

Step 3: Plan and set standards
- Have you set objectives with your employees?
- Have you reviewed accident records to see what general standards of health and safety you have?
- Have you set targets and benchmarks?
- Is there a purchasing policy regarding safety standards required?
- Are there procedures for approving contractors?
- Have safe systems of work been identified?
- Have risk assessments been completed?
- Is the hierarchy of risk control followed?
- Have hazards to persons other than employees been assessed?
- Has a training plan and policy been developed and are there minimum levels of training for all employees?
- Are targets and objectives measurable, achievable and realistic?
- Is there an emergency plan in place?
- Have fire safety procedures been completed?
- Is there a 'zero tolerance' policy with regard to accidents?
- Has the safety consultation process with employees been established?
- Is there a culture of continuous improvement in respect of health and safety?

Step 4: Measure your performance

- Have you or can you measure your health and safety performance
  - where you are
  - where you want to be
  - where the difference is
  - why?
- Are you practising active monitoring or is it just reaction when things go wrong?
- Is there a culture to record near misses or do you wait for accidents to happen?
- Can you benchmark how you perform against another department, company or other organisation?
- Do you know how well you are really doing or does it just look good on paper?
- Are there ongoing accident and incident records and trend analysis?
- Is the effectiveness of training measured – do you assess learning outcomes?
- Is there good legal compliance with health and safety law? Are you up to date with legal changes?

Step 5: Audit and review

- Are you regularly checking that the business is safe and minimising risks to health and safety?
- Is there a formal audit review process?
- Do risk assessments get reviewed proactively?
- Are accidents investigated and processes changed as a result?
- Is there a formal audit process?
- Is it independent?
- Do staff get involved?
- Do you share the findings of the reviews?
- Does the board get kept up to date?
- Is there a health and safety committee?
- Is your business genuinely improving in respect of managing health and safety?

Managing health and safety is no different to managing any aspect of a business, and it should be considered just as important as finance, sales, etc.

The HSE has continuously updated its approach to managing safety so as to ensure that businesses are not overburdened with bureaucracy. Following various reviews of health and safety legislation, the HSE recommended a new approach to managing health and safety based on Plan—Do—Check—Act. This is aligned with the five-step approach described above as shown in Table 2.1.

Table 2.1 Plan—Do—Check—Act and the five-step approach

Plan—Do—Check—Act	Conventional health and safety management	Process safety
Plan	Determine your policy Plan for implementation	Define and communicate acceptable performance and resources needed Identify and assess risks Identify controls Record and maintain process safety knowledge Implement and manage control measures
Do	Profile risks Organise for health and safety Implement your plan	
Check	Measure performance (monitor before events, investigate after events)	Measure and review performance Learn from measurements and findings of investigations
Act	Review performance Act on lessons learned	

Risk Assessments: Questions and Answers
ISBN 978-0-7277-6076-0

ICE Publishing: All rights reserved
http://dx.doi.org/10.1680/raqa.60760.037

publishing

Chapter 3
The Management of Health and Safety at Work Regulations 1999

What is important about the Management of Health and Safety at Work Regulations 1999 (MHSW)?

The Management of Health and Safety at Work (MHSW) Regulations contain specific and goal-setting requirements for health and safety, and cover all employment situations and all types of employment, self-employed persons and others whose undertaking may have a health and safety impact on people.

The MHSW Regulations were introduced in 1992 as part of the health and safety 'six pack' of regulations, implemented in the UK as a result of a number of European Union (EU) directives.

The MHSW Regulations set out specific areas of health and safety management that employers must address, and gave explicit duties to employers and others rather than the then fairly 'wishy-washy' implicit duties imposed by the Health and Safety at Work etc. Act 1974.

The most important aspect of the MHSW Regulations is the requirement for all employers and the self-employed to carry out risk assessments for all work activities undertaken by their employees or others.

The MHSW Regulations were amended in 1999 following various EU directives and instructions to the UK Government that the 1992 Regulations did not fully adopt the principles of the EU Social Charter and its intent in respect of harmonising health and safety law across Europe.

The MHSW Regulations are the backbone of the UK's health and safety management regime and must always be used as the benchmark for health and safety compliance.

Many topic-specific regulations cross-reference to the MHSW Regulations and two or more sets of regulations will need to be read in association.

What is the requirement for risk assessments?

Regulation 3 of the MHSW Regulations sets down the requirement on employers to carry out risk assessments.

The duty is absolute on the employer in that the Regulation states that the employer *shall* make a suitable and sufficient risk assessment.

Regulation 3 requires specifically that employers make a suitable and sufficient assessment of:

- the risks to the health and safety of their employees to which they are exposed while they are at work
- the risks to the health and safety of persons not in their employment arising out of or in conjunction with the conduct by employers of their undertaking.

Regulation 3 also requires self-employed persons to make a suitable and sufficient assessment of:

- the risks to their own health and safety to which they are exposed while at work
- the risks to the health and safety of persons not in their employment arising out of or in connection with the conduct by self-employed persons of their undertaking.

If someone is self-employed in a low-risk environment, such as an office, and no one else works with them, then under regulations introduced in October 2015, health and safety law, and in particular the need for risk assessment, has been removed.

The Health and Safety at Work etc. Act 1974 (General Duties of Self-Employed Persons) (Prescribed Undertakings) Regulations 2015 list specific work environments that are considered high risk, and in these circumstances the legislation on health and safety will still apply to the self-employed, in particular the need to complete risk assessments.

The following industries are considered to be high-risk:

- construction
- agriculture
- working with asbestos
- gas
- railways
- genetically modified organisms (GMO).

What is the legal requirement for reviewing risk assessments?

Regulation 3, sub-section (3) of the MHSW Regulations covers the requirement for the review of a risk assessment and states that:

any assessment such as is referred to in paragraph (1) and (2) above shall be reviewed by the employer or self-employed person who made it if:

- there is reason to suspect that it is no longer valid; or
- there has been a significant change in the matters to which it relates; and
- where as a result of such a review changes to an assessment are necessary, the employer or self-employed person shall make them.

What is the legal requirement regarding the employment of young persons?

Before a young person is employed the employer must carry out a specific risk assessment that complies with Regulation 3 of the MHSW Regulations 1999.

Employers must take particular account of:

- the inexperience, lack of awareness of risks and immaturity of the young person
- the fitting out and layout of the workplace and work station
- the nature, degree and duration of exposure to physical, biological and chemical agents
- the form, range and use of work equipment and the way in which it is handled
- the organisation of processes and activities
- the extent of health and safety training provided or to be provided to the young person
- the risks from agents, processes and work.

Employers must record the significant findings of any assessment in writing if they have five or more employees.

What is the law on the 'principles of prevention'?

Regulation 4 of the MHSW Regulations 1999 states that where employers implement any preventive and protective measures they shall do so in accordance with the following principles:

- avoiding the risks
- evaluating the risks that cannot be avoided
- combating the risks at source
- adapting the work to the individual, especially with regard to the design of workplaces, the choice of work equipment and the choice of working and

production methods, with a view, in particular, to alleviating monotonous work and work at a predetermined work rate, and to reducing their effect on health
- adapting to technical progress
- replacing the dangerous with the non-dangerous or the less dangerous
- developing a coherent, overall prevention policy that covers technology, organisation of work, working conditions, social relationships and the influence of factors relating to the working environment
- giving collective protective measures priority over individual protective measures
- giving appropriate instructions to employees.

A court of law would expect employers facing a prosecution for inadequate controls of risk to be able to demonstrate that they have followed the principles of prevention and protection.

What are an employer's duties in respect of health surveillance?
All employers shall ensure their employees are provided with such health surveillance as is appropriate, having regard to the risks to their employees' health and safety which are identified in the risk assessment.

Health surveillance is required under the Control of Substances Hazardous to Health Regulations 2002 and also under the MHSW Regulations 1999.

If a risk assessment shows:

- that there is an identifiable disease or adverse health condition related to the work concerned
- valid techniques are available to detect indications of the disease or condition
- there is a reasonable likelihood that the disease or condition may occur under the particular conditions of work
- surveillance is likely to further protect the health and safety of employees

then health surveillance should be introduced.

A competent person must determine the level and frequency of health surveillance.

The primary benefit of health surveillance is to detect adverse conditions at an early stage, thereby enabling further harm to be prevented.

It may be necessary to consult with the Employment Medical Advisory Service if, as an employer, you do not have a competent person to assist with health surveillance assessments.

There are some specific regulations that require an employer to assess the need for, and then offer, health surveillance, and these need to be considered in addition to the general duties under the MHSW Regulations 1999.

- Control of Substances Hazardous to Health Regulations 2002
- Control of Lead at Work Regulations 2002
- Control of Asbestos at Work Regulations 2012
- Noise at Work Regulations 2005

Under the above regulations medical examinations may also be required.

As an employer, can I deal with all health and safety matters myself?

Yes, provided you could convince the enforcement authorities, and probably the courts, that you are competent to deal with health and safety matters.

Regulation 7 of the MHSW Regulations 1999 requires employers to appoint one or more 'competent' persons to assist them in undertaking the measures they need to take to comply with the requirements and prohibitions imposed upon them by virtue of the relevant statutory provisions.

What do the MHSW Regulations 1999 require regarding competent persons?

Many aspects of health and safety merely require good practical common sense. 'Competency', however, is described as having sufficient training and experience or knowledge and other qualities to enable a person to properly assist in undertaking the measures needed to comply with statutory duties.

Any or all competent persons appointed must cooperate with one another.

The employer must ensure that any competent person is provided with relevant information about hazards and risks, etc.

Where practicable, and where available, a competent person should be appointed from within the workforce; the appointment should not be external unless specialised knowledge and experience is required.

What are the requirements regarding 'serious and imminent' danger?

Employers must adopt appropriate procedures for dealing with situations that cause serious and imminent danger to persons at work.

Competent persons must be nominated in sufficient numbers to implement the emergency plans – in particular, any evacuation of people from the workplace.

Employees must be made aware of the hazards that could cause an imminent or serious situation to arise, and must be informed of the procedures to be followed to protect themselves from the danger.

In any situation that creates serious or imminent danger to employees, there must be a procedure that enables them to evacuate the area safely and to proceed to a place of safety. It should be accepted that work processes will stop immediately.

Emergency procedures must be written down and available for all employees and others.

A risk assessment should determine what emergency procedures will be necessary, and the employer should think about more than just fire and bomb procedures.

Examples of emergency procedures are:

- fire
- bomb
- explosion
- chemical release
- flood
- toxic gas release, fumes, etc.
- unexpected shutdown of exhaust ventilation, which could cause toxic fumes, dust, etc. to build up
- terrorist attack
- radiation leak
- biological agent release.

Employers need to consider the possibilities of the above and plan for the event by considering what they would do, who would do what, how people will get out, where will they go, how they will be accounted for, etc.

What is required regarding 'contact with emergency services'?

Employers need to assess the need for, and likelihood of, contacting the emergency services.

Often, emergency services are not familiar with the hazards associated with an employer's undertakings, and emergency personnel can be placed at great health and safety risk.

Regulation 9 of the MHSW Regulations 1999 requires a formal arrangement to be made so that the emergency services have some knowledge of any special safety hazards, or they have been informed if the employer believes that there is a risk of the occurrence being a 'serious and imminent danger'.

Records should be kept of any contact, meetings, reviews, etc. with the emergency services.

It may be appropriate to send a copy of the emergency plan to the authorities for their fire records.

For many employers, it will be sufficient to ensure that employees know the emergency services' phone number and how to contact them for assistance.

What is the legal duty under the MHSW Regulations 1999 for information to be given to employees?

Regulation 10 governs the provision of information to employees and requires that employers provide employees with 'comprehensible and relevant' information on:

- the risks to their heath identified by the risk assessments
- preventive and protective measures
- emergency procedures
- fire safety arrangements
- the identity of nominated persons who will take charge of emergency situations (e.g. fire wardens)
- any risks notified to the employer as being present on another employer's premises and to which the employee may be exposed.

When employers propose to employ a young person they shall inform that person's parents or guardians of the risk assessment, control measures, etc. to be adopted so as to ensure that the young person is kept safe while at work.

What happens in relation to health and safety when two or more employers share a workplace?

Health and safety law requires that two or more employers who share a workplace must cooperate with one another in respect of health and safety matters and, where necessary, appoint a coordinator for health and safety.

Employers must take reasonable steps to reduce risks both to their own employees and to others, and must not put other people at undue risk.

Often, a multi-occupied building will have a managing agent, and it would be for this person, as the person in control of the premises, to coordinate emergency procedures, etc. and to undertake risk assessments for the 'common parts'.

Employers have a duty to inform those who need to know about the hazards and risks associated with their undertaking, and employers must consider the risks posed by a different employer when determining their own risk assessments.

It is good practice to swap health and safety policies with other occupiers so that each is fully aware of the hazards, risks and control measures posed by each employer.

Emergency procedures (e.g. evacuation) need to be coordinated so that there is a unified response to any alarm.

Where employees are working in a host employer's premises (e.g. maintenance workers), the host employer must ensure that the visiting employees are provided with relevant information on hazards, risks, control measures and emergency procedures.

In particular, the names of nominated persons in respect of emergencies should be made available to visiting employees (e.g. fire warden and first aiders).

Common practice for sharing information on safety matters for visiting employees include:

- signing in and an induction briefing
- contractors' handbook
- pre-approval of contractors
- joint training sessions
- permit to work systems.

What are employers' duties regarding the provision of training for their employees?

Regulation 13 of the MHSW Regulations 1999 requires employers to provide their employees with adequate training in respect of health and safety:

- on being recruited
- on being exposed to new or increased risks due to:
  - being transferred to another department or section
  - the provision of new equipment
  - the change of existing equipment
  - the introduction of new technology
  - the introduction of a new system of work.

Training shall be repeated periodically and shall take place during working hours. It shall also be free of charge.

The provision of information, instruction and training must be comprehensible and understood by employees.

For the purposes of health and safety, temporary workers must be treated as permanent employees and must also receive information, instruction and training.

What duties do employees have under the MHSW Regulations 1999?

Regulation 14 states the requirements for employees as follows:

Every employee shall use any machinery, equipment, dangerous substance, transport equipment, means of production or safety device provided to him by his employer in accordance both with any training in the use of the equipment concerned which has been received by him and the instructions respecting that use which have been provided to him by the said employer in compliance with the requirements and prohibitions imposed upon that employer by or under the relevant statutory provisions.

Every employee shall inform his employer or any other employee of that employer with specific responsibility for the health and safety of his fellow employees

(a) of any work situation which a person with the first-mentioned employee's training and instruction would reasonably consider represented a serious and immediate danger to health and safety; and

(b) of any matter which a person with the first-mentioned employee's training and instruction would reasonably consider represented a shortcoming in the employer's protection arrangements for health and safety

in so far as that situation or matter either affects the health and safety of that first mentioned employee or arises out of or in connection with his own activities at work, and has not previously been reported to his employer or to any other employee of that employer in accordance with this paragraph.

What responsibility do employers have for temporary workers?

Regulation 15 of the MHSW Regulations 1999 states that:

Every employer shall provide any person whom he has employed under a fixed-term contract of employment with comprehensible information on

(a) any special occupational qualifications or skills required to be held by that employee if he is to carry out his work safely; and

(b) any health surveillance required to be provided to that employee by or under any of the relevant statutory provisions

and shall provide the said information before the employee concerned commences his duties.

Every employer and every self-employed person shall provide any person employed in an employment business who is to carry out work in his undertaking with comprehensible information on

(a) any special occupational qualifications or skills required to be held by that employee if he is to carry out his work safely; and
(b) health surveillance required to be provided to that employee by or under any of the relevant statutory provisions.

Every employer and every self-employed person shall ensure that every person carrying on an employment business whose employees are to carry out work in his undertaking is provided with comprehensible information on

(a) any special occupational qualifications or skills required to be held by those employees if they are to carry out their work safely; and
(b) the specific features of the jobs to be filled by those employees (in so far as those features are likely to affect their health and safety)

and the person carrying on the employment business concerned shall ensure that the information so provided is given to the said employees.

Case study

Hugo Boss

At Oxford Crown Court in September 2015 Judge Peter Ross fined the company £1.1 million for breaching Section 3(1) of the Health and Safety at Work etc. Act 1974 and a further £100 000 under the Management of Health and Safety Work Regulations 1999. The company also had to pay almost £47 000 costs.

Austen Harrison, 4 years old, died when a 250 lb wall mirror fell on him in the changing room area of the shop at the discount outlet village. Barry Berlin, prosecuting on behalf of Cherwell District Council, told the court that the 2.1 m tall, three-way mirror should have been fixed to a reinforced wall after the store was refurbished in summer 2012.

The prosecution said the company had experienced 'systemic failures' in their implementation of health and safety checks at the store. Mr Berlin said that Simon Harrison, Austen's father, had gone into the changing area – where there was a three-way mirror weighing 18.45 stone – to try on a suit.

'Unknown to the Harrison family at that time, and it seems unrecognised by anyone at Hugo Boss, that mirror had not been fixed to the wall but had negligently been left free standing without any fixings,' Mr Berlin said. 'While Simon Harrison was trying on the suit, Austen was moving the wings of the mirror.'

Instructions for the mirror state that it should be properly affixed to a reinforced wall. However, the mirror was standing against a stud wall. Mr Berlin said contractors had 'hurried' to convert the pop-up shop from a Burberry store that had been in the space previously.

The company also settled a civil claim for damages brought by the boy's family.

Risk Assessments: Questions and Answers
ISBN 978-0-7277-6076-0

ICE Publishing: All rights reserved
http://dx.doi.org/10.1680/raqa.60760.049

Chapter 4
Undertaking risk assessments

Who should undertake a risk assessment?

Regulation 3 of the Management of Health and Safety at Work (MHSW) Regulations 1999 states that:

> every employer shall make a suitable and sufficient assessment of:
>
> (a) the risks of the health and safety of his employees to which they are exposed whilst they are at work; and
> (b) the risks to the health and safety of persons not in his employment arising out or in connection with the conduct by him of his undertaking
>
> for the purposes of identifying the measures he needs to take to comply with the requirements or prohibitions imposed upon him by or under the relevant Statutory Provisions.

Consequently, the law requires an *employer* to carry out a risk assessment.

Under the terms of the organisation's safety policy, the employer can delegate the responsibility for undertaking risk assessments to others (e.g. safety officers and departmental managers).

With the exception that the employer carries ultimate responsibility for the risk assessment, it can be conducted by anyone authorised to do so.

In many organisations, the best people to carry out a risk assessment on job tasks are the employees. This is because they are familiar with the hazards and risks of what they do and they know the actual way they carry out the tasks, as opposed to the 'theoretical way'.

The law requires that anyone involved in health and safety matters for an employer, including the employers themselves, must be competent to do so.

Regulation 7 of the MHSW Regulations 1999 requires employers to appoint competent persons to assist in health and safety matters. This will include a competent person being appointed to undertake risk assessments.

Regulation 7 defines persons as being competent if they have sufficient information, knowledge and experience to enable them to properly assist the employer in discharging his responsibilities.

Do self-employed people have to carry out risk assessments?

If self-employed people employ others they will need to complete risk assessments to ensure that those employees are not put at risk.

If self-employed people are working for themselves, providing goods and services to others, they will not be required to complete risk assessments unless they are working in high-risk environments or carrying out high-risk tasks.

If the tasks carried out by a self-employed person pose health and safety risks to others then the self-employed person must complete risk assessments to ensure that the safety of others is protected.

October 2015 saw the introduction of the Health and Safety at Work etc. Act 1974 (General Duties of Self-Employed Persons) (Prescribed Undertakings) Regulations 2015 as a result of a Government review of health and safety law. These regulations simplify health and safety law for the self-employed, provided they are not working in the following areas:

- agriculture
- construction
- asbestos
- gas
- genetically modified organisms (GMO)
- railways.

Does absolutely every single job activity require a risk assessment?

No, although it sometimes feels like that!

The requirement of the MHSW Regulations 1999 are as follows:

Every employer shall make a suitable and sufficient assessment of

(a) the risks to health and safety of his employees to which they are exposed while they are at work, and

(b) the risks to the health and safety of persons not in his employment arising out of or in connection with the conduct by him of his undertaking for the purpose of identifying the measures he needs to take to comply with the requirements and/or prohibitions imposed upon him under the relevant statutory provisions.

If a work activity does not pose any health and safety risks then there is no need to carry out a risk assessment, although a risk assessment of sorts will be carried out in order to establish that the job task has no hazards or risk attached to it.

Increasingly, however, it is best practice to undertake risk assessments for all job tasks because, even though the statutory laws may not require them, the need to provide a duty of care under civil law makes risk assessments a valuable defence tool.

Do risk assessments have to be in writing?

If you have less than *five* employees, risk assessments do not need to be written down, although it is always good practice to do so as you never know when you might need to produce a record.

Where an employer has *five* or more employees all risk assessments must be written down.

Do risk assessments have to be signed by employers and employees?

No, there is no legal requirement for risk assessments to be signed.

You may wish to implement a procedure where risk assessments are signed once an employee or other person has been instructed on the content of the risk assessment, as this will provide you with suitable and sufficient training records and evidence that employees and others were aware of the hazards and risks associated with the tasks they are undertaking.

Is there a standard format for a risk assessment?

No, mainly because risk assessments are individual to job tasks and need to be 'site specific'!

The Health and Safety Executive (HSE) publishes guidance on how to complete risk assessments and a risk assessment template.

There is no right or wrong way to complete a risk assessment. The law requires that it is 'suitable and sufficient'.

A risk assessment must contain suitable information that is useful to employees to understand what hazards they may be exposed to when carrying out the task.

Generally, any format that includes the following will be suitable:

- description of the job task
- location of activity
- who will carry out the activity
- who else might be affected by the task
- the hazards that have been identified
- what could go wrong
- what the injuries might be and how severe they might be
- how likely the risks are
- what can be done to reduce or eliminate the hazards
- the information that employees or others need to work safely
- when the risk assessment might be reviewed.

What do the terms 'hazard' and 'risk' mean?

A hazard is something with the potential to cause harm.

The risk is the likelihood that the potential harm from the hazard will be realised.

The extent of the risk will depend on:

- the likelihood of the harm occurring
- the potential severity of that harm (resultant injury or adverse health effect)
- the extent of people who might be affected (i.e. several people, vast groups or communities at large (e.g. from chemical releases)).

What does 'suitable and sufficient' mean in respect of risk assessment?

The phrase 'suitable and sufficient' is not defined in the MHSW Regulations 1999 or in the Health and Safety at Work etc. Act 1974.

The Management of Health and Safety at Work Code of Practice (ACOP – L21) states that:

> The level of risk arising from the work activity should determine the degree of sophistication of the risk assessment.

Insignificant risks can generally be ignored, as can routine activities associated with life in general.

Risk assessments are expected to be proportionate to the hazards and risks identified.

Enforcement officers do not expect to see huge volumes of paperwork – the simpler the risk assessment and the clearer the information, the easier the employee will find it to follow safe procedures.

When should risk assessments be reviewed?

The MHSW Regulations 1999 require a risk assessment to be reviewed if

- there is reason to suggest that it is no longer relevant or valid
- there has been significant change in the matters to which it relates.

If changes to a risk assessment are required, the employer has a duty to make the changes and reissue the risk assessment.

Employers are not expected to anticipate risks that are not foreseeable.

However, if events happen that alter information available or the perception of risk, the employer will be expected to respond to the new information and assess the hazards and risks in the light of the increased knowledge.

Accidents and near misses should be investigated, as these incidents will indicate whether more knowledge is available on the hazard or risk associated with the job. The risk assessment may need to be reviewed because:

- something previously unforeseen has occurred
- the risk of something happening or the consequences of the event may be greater than expected or anticipated
- precautions prove less effective than anticipated.

New equipment, a new working environment, new materials, different systems of work, etc. will all require existing risk assessments to be reviewed.

How long do I need to keep my risk assessments for?

There is no set amount of time for which you need to keep your records relating to general risk assessment. It is good practice, however, to keep them while they remain relevant.

I share my workplace with another employer. How should we manage risk assessments?

If you share a workplace with another employer or a self-employed person you will both need to:

- tell each other about the specific risks in your business that may affect the other employer
- cooperate and coordinate with each other to control the health and safety risks.

It would be sensible to set up regular meetings to discuss work activities and any that would overlap and to ensure that a proper plan is in place to manage hazards and risks affecting all persons and outlining the ways such information will be communicated to those who need to know.

There may be areas in the premises that require a combined approach to risk assessment, and each employer should agree who is to complete the 'common parts' risk assessment and how this will be shared with the other.

Which people must be considered as being exposed to the risks from work activities?

- Employees of the employer
- young workers and those on work experience
- new and expectant mothers
- cleaners – whether contract or in-house
- visitors to the premises
- maintenance workers – both contract and in-house
- members of the public
- employees of other employers with whom you share the building or premises
- delivery drivers
- sales representatives
- people with disabilities – extra control measures may be needed to protect them from risk
- peripatetic workers – those working away from the office, usually visiting other workplaces or people's homes (e.g. midwives)
- volunteers.

Must risk assessments be categorised into high, medium or low risks?

Not necessarily by law, but it is good practice to identify the extent of harm that an employee or other person could be exposed to.

Even after all precautions have been taken some risk (i.e. potential cause of injury or ill health) may remain. This is often referred to as 'residual risk'

Residual risk is either 'high, medium or low', or 'very likely, probable or unlikely'.

Some risk management approaches allocate numerical scores to various types of risk and the severity of those risks. Multiplying one score by the other gives a 'risk rating'. Scores above a set target become unacceptable and measures must be put in place to reduce the risks.

What are some of the common control measures that can be put in place to reduce the risks from job activities?

The aim of risk assessment is to reduce the residual risk associated with a task to as small as possible.

First, try to eliminate the hazard altogether – why do something or use something if you do not have to?

Where the hazard cannot be eliminated it must be reduced to an acceptable level by implementing *control measures*.

A common approach is to follow the 'hierarchy of risk control', namely:

- try a less risky option (i.e. substitute something less hazardous)
- prevent access to the hazard (e.g. guarding)
- organise work to prevent exposure to the hazard
- issue personal protective equipment
- protect the workforce as a whole (e.g. through exhaust ventilation)
- provide welfare facilities to aid removal of contamination, to take rest breaks, etc.
- provide first-aid facilities.

What will an enforcement officer expect from my risk assessments?

Enforcement officers will want to see that you have:

- completed risk assessments
- considered site-specific issues
- completed comprehensive checks of the workplace
- involved workers
- considered the hazards and risks to others
- dealt with immediate hazards to reduce risks
- a system for reviewing risk assessments
- suitable records
- provided information, instruction and training to your employees
- introduced suitable and sufficient measures.

Site-specific risk assessments are probably the most important aspect. Environmental health officers and HSE inspectors are not keen on 'generic' risk assessments unless steps have been taken to ensure that any special site hazards and precautions have been added to the risk assessment.

To enforcement officers, risk assessment is not a paper exercise to be pulled off the shelf in a manual. It is a proactive approach by employers to consider what could harm their employees and others and what measures they intend to take to reduce the risks of injury and ill health.

What is meant by the term 'control measure'?

A control measure is the precautions deemed necessary to reduce the consequences of the hazard and risk (i.e. to reduce the risk to an acceptable level).

A control measure could be:

- physical (e.g. guard to a machine)
- substitution (e.g. use a less hazardous substance)
- a system of work
- personal protective equipment
- separation of the worker from the environment or machine
- environmental controls (e.g. ventilation).

The risk assessment identifies what control measures or precautions are necessary in order to manage the risks that have been identified from the hazard being assessed.

What is the 'hierarchy of risk control'?

When a hazard has been identified, the most effective way of reducing the effect of the hazard, or the risk, is by eliminating the hazard completely. Unfortunately, this is not always possible and a staged approach to controlling the risk has to be adopted.

This staged approach to controlling risk is called the 'hierarchy of risk control'.

1 The first, and most effective, stage is to deal with a hazard and its risk in order to eliminate the hazard completely. No hazards = no risks.
2 If stage 1 cannot be achieved, substitute the identified hazard for a less harmful hazard. For example, why use a substance that can cause cancer when an alternative is on the market that may only be a minor skin irritant, or why carry 50 kg bags of cement when 25 kg bags are available and easier to carry?
3 If a suitable substitute cannot be found and the original hazard has to remain, protect the workforce as a whole from the hazard (e.g. increase the ventilation in the workshop so that the hazard is tackled at source).

4   If stage 3 is not possible, provide all employees with individual worker protective clothing and equipment so that their own individual health is protected (e.g. masks, goggles and local exhaust ventilation).
5   Finally, monitor and review the controls that you have put in place to make sure that they are effective, otherwise it will be back to stage 1 – eliminating the hazard.

What is the common approach to risk assessment?

The HSE has pioneered a *five-step approach* to risk assessment (the original document outlining the steps has been replaced with *Risk Assessment: A Brief Guide to Controlling Risks in the Workplace*, INDG163(rev 4), 2014).

The five steps are:

Step 1:   Look for the hazards
Step 2:   Decide who might be harmed and how
Step 3:   Evaluate the risks and decide whether the existing precautions are adequate or whether more should be done
Step 4:   Record your findings
Step 5:   Review the assessment and revise it whenever necessary.

What needs to be considered in each of the five steps

First, risk assessment should not be overcomplicated. Remember, risk assessment is a careful examination of what, in your work, could cause harm to people, and an assessment of what you need to do to prevent harm to people. You may already have taken enough precautions to protect them or you may need to implement some more.

You will need to decide whether a hazard is significant and whether you have covered it by undertaking satisfactory precautions so that the risk is small.

Step 1: Look for the hazards

Walk around the workplace and look out for anything that could cause harm. Take a fresh view and do not make assumptions about anything.

Concentrate on identifying the serious hazards that could cause major harm to people (i.e. cause significant injury or affect several people).

Talk to your employees and ask them what they think of as hazards. Ask them how they really do the job and whether they follow the rules or do the job a little differently.

Review any records that you have such as those on accidents and incidents. It is a good idea to keep 'near miss' records because these should tell you that something is not working as it should but that things have not got so bad as to cause an accident.

What type of accidents are happening and why? For instance, are people off sick for periods of time with back injury?

Check manufacturers' operating procedures, manuals and instructions.

Check what substances are being used and what substances are being produced by the works process itself (e.g. dust).

Common hazards:

- equipment – how it is used, guards, controls, noise
- work processes – how things are done, systems to be followed
- environmental conditions – condition of floors, heating, ventilation, etc.
- materials in use – chemicals, gases, substances.

Step 2: Decide who might be harmed and how
Consider absolutely everybody who could be harmed by the hazards you have identified.

Remember to include:

- all employees
- agency staff
- self-employed people
- visitors, public, etc.
- contractors
- cleaners
- delivery personnel
- maintenance workers.

How might they be affected by the hazard? Are they likely to receive a serious injury or none at all? Could they be harmed because they were in the vicinity of the hazard or because they have to undertake the job task?

Step 3: Evaluate the risks and decide whether existing precautions are adequate or more should be done
Consider how likely it is that each hazard you have identified could cause harm.

Then consider what you are currently doing to reduce the potential for harm.

Are you doing enough? Have you done all the things that the law requires you to do? Are you following good practice and industry standards? Are manufacturers' instructions being adhered to? If, for instance, you use dangerous machinery, have you got guards in place so that the dangerous parts cannot be accessed?

What more could be done to reduce the risks to as low a level as possible?

Example

You may have identified the hazard of working at heights – there is definitely a potential harm to people from falling. You may have control measures in place because you have provided scaffolding, handrails and toeboards as required by the Work at Height Regulations 2005. But have you done enough? Have you considered whether your employees or others *need* to work at height in the first place? Could they have a safer means of access to the place of work than the scaffolding? In itself a scaffold is dangerous or hazardous. Could mobile elevating work platforms be used instead?

If you are not satisfied that all risks are as small as possible then more needs to be done.

Draw up an action plan so that you work on eliminating the highest risks first, or those which could harm most people.

If you need to provide more control measures or precautions, consider the hierarchy of risk control:

- eliminate the hazard
- substitute for a less risky option
- prevent access to the hazard
- organise the work to reduce exposure to the hazard
- issue personal protective equipment
- provide welfare facilities (e.g. washing and first aid facilities).

Consider all types of work environments in which employees work and whether they work in someone else's premises. Employers must assess the hazards and risks to their employees, and even though you may not know what these are, you are responsible if your employee works somewhere else. So, ask the owner of that building or business what *their* hazards and risks are, and assess those in relation to the job you expect your employee to do.

Persons in control of premises have duties under health and safety even if they are not employees and the MHSW Regulations 1999 cover the cooperation of employers and others where there is a multi-occupied site.

Step 4: Record your findings

If you have five or more employees, you need to record the *significant findings* of your risk assessment process in writing.

Small companies with fewer than five employees are exempt from this, although it is always good practice to keep some records, as you never know when you might want to prove what you have done.

As stated earlier, there is no standard risk assessment form or template for records.

You will need to decide what type of form is appropriate for record keeping. The HSE gives some guidance on a simple layout for a risk assessment, but there is no right or wrong way to go about it.

Remember that your risk assessment must be 'suitable and sufficient' – this does *not* mean perfect!

The records you keep need to show that you have considered the hazards, identified the people at risk, determined and actioned the control measures necessary to reduce the risks, and considered when the risk assessment needs to be reviewed.

Step 5: Review the assessment and revise it whenever necessary

Risk assessment is not a 'once and for all' exercise. Hazards in any workplace may change, the circumstances in which those hazards occur may change, the people may change, and the materials and equipment may change.

So, a risk assessment needs to be constantly reviewed to see if it is still relevant.

Your original risk assessment process should identify when and if hazards and risks will change, and should indicate a regular review period.

If things stay fairly static, a risk assessment with a low residual risk may only need to be reviewed annually. But a works process or job task with a high residual risk and which relies on effective control measures to make the risk tolerable will need to be reviewed much more regularly.

What are site-specific risk assessments?

Site-specific risk assessments review the hazards and risks associated with an actual job on a specific site or within specific premises.

The law is concerned about what may *actually* happen to an employee or other person while they are at work, or affected by its activities, not what could happen because the employer has brainstormed every conceivable hazard in every conceivable location.

Hazards may be quite common across a range of work activities, and there is a tendency for employers to produce 'generic risk assessments'. These are records of common hazards and risks but they do not address the actual work environment.

Generic risk assessments do have a valuable part to play in the process of risk assessment but they need to be reviewed in the light of what actually happens in the workplace.

Are generic risk assessments acceptable under health and safety law?

There is no law against generic risk assessments produced by, for example, a trade body being used as the employer's risk assessment.

However, the question to consider is whether the generic risk assessment is 'suitable and sufficient' for the purposes of the MHSW Regulations 1999.

Prosecutions can be taken by the enforcing authorities for inadequate risk assessments, and this is just as serious an offence as having *no* risk assessments.

An enforcement authority is more likely to serve an improvement notice on an employer for failing to have a suitable and sufficient risk assessment, and will require improvements to be made. Alternatively, inspectors who are concerned about the risk from a job task or activity could serve a prohibition notice under the Health and Safety at Work etc. Act 1974.

As an employer, I have completed my risk assessments, recorded the significant findings in writing and have a risk assessment manual. What else do I need to do?

The law requires you to tell your employees and others about the findings of your risk assessment exercise.

Employees must know what hazards they are exposed to during their working day and must be advised of the precautions or controls that have been implemented to help reduce the risks of harm.

Employees could be given copies of the risk assessments individually as part of their employee handbook, or they could be given a formal training session that identifies the hazards and risks and trains them in how to operate or follow the control measures.

Information could be displayed on company notice boards or adjacent to the work areas.

As long as employees are kept informed, the method by which it is done is left to the employer.

Does an employer have to carry out separate risk assessments for employees with disabilities?

No, there is no requirement to carry out a separate risk assessment for a disabled employee. Employers should already be managing any significant workplace risks, including putting control measures in place to eliminate or reduce the risks. An employer who becomes aware of an employee who has a disability should review the risk assessment to make sure it covers risks that might be present for that particular employee.

How does an employer assess whether an employee's disability puts either themselves or others at risk?

An employer must not make assumptions about an employee's condition and should consider a number of things.

- Have the risks to all employees been assessed properly and appropriate control measures put in place?
  It may be that appropriate changes to the work equipment and environment could significantly reduce the risk and take the issue of disability out of the equation.
- Does the employee's condition create an increased risk to their own health and safety or the health and safety of others?
  The employee's condition might be well managed and the employer may conclude that review of the situation at regular intervals is sufficient.
- If the employee's condition does create an increased risk to their own health and safety or the health and safety of others, can these risks be prevented or adequately controlled through normal health and safety management?
  Can the risks be addressed by allowing other colleagues to do certain elements of the activity, by providing suitable alternative equipment (e.g. automated equipment to reduce manual handling) or by changing systems of work?
- If not, what reasonable adjustments could be put in place to prevent or adequately control the residual risks?
  The employer might be able to apply for financial assistance through the Government's Access to Work scheme to cover the cost of new equipment.
- Be sure to consult with the employee themselves and colleagues to seek opinions, and ensure they are involved in discussions that affect them. Those involved in the work often propose good solutions.

Risk Assessments: Questions and Answers
ISBN 978-0-7277-6076-0

ICE Publishing: All rights reserved
http://dx.doi.org/10.1680/raqa.60760.063

Chapter 5
Control of substances hazardous to health and dangerous and explosive substances

What are the COSHH Regulations 2002?

The Control of Substances Hazardous to Health Regulations 2002 are known as COSHH.

The COSHH Regulations 2002 set out the duties that employers have to their employees and others to protect them from exposure and harm from hazardous substances.

The 2002 regulations came into force in November 2002 and replaced all earlier sets of regulations (i.e. 1988, 1994 and 1999).

The regulations were amended in March 2003 to address further issues in respect of carcinogens.

What do the COSHH Regulations 2002 require?

Five basic principles of occupational hygiene underline the COSHH Regulations 2002:

- identify the hazardous substance, identify how it is to be used, and assess the risk to health, precautions and health risks arising from that substance
- if the substance is harmful, wherever possible substitute it for a less harmful substance
- introduce appropriate measures to prevent or control risks, and ensure that control measures are used, that any protective equipment is properly maintained and that any safety procedures are observed
- where necessary, monitor the exposure of employees and introduce an appropriate form of surveillance of their health
- inform, instruct and train employees in the risks to their health and safety and the precautions that need to be taken.

What substances are covered by the COSHH Regulations 2002?

The regulations cover a wide range of substances and includes those that are very toxic, harmful, corrosive, irritant or biological in nature. Such substances could include cleaning materials for floors, toilets and drains, glasswasher and dishwasher detergents, pest control materials, dusts, fumes, solvents, building products, oils, etc.

There is no limit on the quantity of chemicals with regard to the application of the COSHH Regulations 2002. The overriding principle is that if a substance is a hazard to health it *must* be assessed.

All substances are safe when properly used, but the use of each must be assessed. Employees must be made aware of any hazards and the precautions necessary, and be trained in how to use the substances correctly.

What changes did the COSHH Regulations 2002 introduce?

The COSHH Regulations 2002 did not fundamentally change employers' duties to ensure that employees and others are not exposed to the harmful effects of hazardous substances.

The regulations generally made changes as follows:

- To numerous definitions within the regulations (e.g. biological agents, and inhalable and respirable dust).
- COSHH assessments under Regulation 6 have been amended to:
    - an employer shall not carry out work that is liable to expose any employees to any substance hazardous to health unless they have
        (a) made a suitable and sufficient assessment of the risk created by that work to the health of those employees and of the steps that need to be taken to meet the requirements of these Regulations
        (b) implemented the steps referred to in sub-paragraph (a).
    - the assessment is to consider:
        o the hazardous properties of the substance
        o information on health effects provided by the supplier, including information contained in the safety data sheet
        o the level, type and duration of exposure
        o the circumstances of the work, including the amount of substance involved
        o activities, such as maintenance, where there is potential for a high level of exposure
        o any relevant occupational exposure limit/standard, maximum exposure limit or similar occupational exposure limit

o the effect of preventive and control measures that have been or will be taken to comply with Regulation 7

o the results of relevant health surveillance

o the results of any monitoring of exposure

o the risks of exposure to more than one substance (i.e. the 'cocktail' effect)

o the approved classification of any biological agent

o such additional information as the employer may need to complete the assessment.

■ The assessment must be reviewed if the results of monitoring show it to be necessary.

■ Employers who employ five or more employees must record the significant findings of the assessment as soon as is practicable after the risk assessment has been made and steps must be taken to implement control measures.

■ There is a specific requirement under Regulation 7 to substitute a substance or process if this eliminates or reduces risks to health.

■ Control measures are listed in order of priority in Regulation 7.

■ Biological agents are now covered in the body of the regulations.

■ All equipment used as part of control measures is to be kept clean.

■ New provisions are made regarding employee monitoring, the keeping of records and health surveillance.

■ Information, instruction and training requirements have been extended to include details on occupational exposure limits, access to relevant safety data sheets, exposure and health risks of the substance, significant findings of the COSHH assessment, results of health surveillance, control measures to be implemented, etc.

Duties extend to training persons other than the employer's employees if those people are exposed to the risks from hazardous substances.

How do dangerous chemical products get into the body?

There are three main ways in which products get into the body: through ingestion, through the skin or through inhalation. The form of the product plays an important role. The more finely divided a product is the more easily it is absorbed (generally, the smaller the particles the more dangerous they are). For example, solids may be in the form of a powder and liquids may be in the form of an aerosol.

Absorption is dependent on many factors, including the state of subdivision of the product (i.e. the smallness of the particles), its concentration, the length of exposure, the use of protective equipment, the solubility of the substance in fat, etc.

A chemical that enters the body via any route can be transported to other parts of the body in the bloodstream and can cause damage to organs.

Digestive route (entry via the mouth)

Entry via the digestive route (or ingestion) is usually accidental or the result of carelessness, for example:

- through transferring a product from one container to another by sucking it up through a pipette, or through a product having been stored in a food or drink container
- through eating, smoking, drinking, etc. after having handled a dangerous product and not having washed hands.

Percutaneous route (entry via the skin)

Certain products, such as irritant and corrosive products, act locally at the place where they come into contact with the skin, the mucous membrane or the eyes.

Others, which are soluble in fat, not only act on the skin but also penetrate it and spread throughout the body, where they can cause various disorders. This is the case with solvents, which degrease the skin, but which can also damage the liver, nervous system or kidneys. Benzene can damage the bone marrow. Motor fuel (which has a relatively high benzene content) should not be used to wash hands.

Small cuts and grazes provide an easy route for dangerous chemicals.

Respiratory route (entry via the lungs)

This is the most common entry route at work, as pollutants can be present in the atmosphere and then enter the lungs with the air breathed in. This can occur, for example, when handling solvents, paints or glues, stripping leaded paint with a blowtorch or welding.

Once inhaled into the lungs, these chemicals enter the bloodstream and can cause damage not only to the respiratory system but also to the rest of the body.

What are workplace exposure limits?

Workplace exposure limits (WELs) are British occupational exposure limits that have been set in order to help protect the health of workers. WELs are concentrations of hazardous substances in the air, averaged over a specified period of time, referred to as a time-weighted average (TWA). Two time periods are used:

Long term – 8 hours
Short term – 15 minutes

Short-term exposure limits (STELs) are set to help prevent effects such as eye irritation, which may occur following exposure for a few minutes.

WELs are listed in a Health and Safety Executive (HSE) document called EH40/2005 *Workplace Exposure Limits*, and it is recommended that employers consult this document in order to establish whether specific exposure limits exist for certain substances. However, the absence of a substance from the list does not indicate that it is safe.

What are occupational exposure limits or standards (OEL/OES)?

For a number of commonly used hazardous substances, the Health and Safety Commission has assigned occupational exposure limits (OELs) or occupational exposure standards (OESs) to help define what is adequate control.

OELs are set at levels which will not damage the health of employees exposed to the substance by inhalation, day after day.

Where a substance has an OEL, the exposure of employees to the substance must legally be reduced to the OEL level.

What are maximum exposure limits?

Maximum exposure limits (MELs) are set for substances that can cause the maximum amount of health damage. These substances usually cause life-threatening illnesses such as cancer, asthma, severe industrial dermatitis or respiratory conditions.

Substances that have an MEL must be used only if there is no alternative, and the exposure time must not exceed the stated limit – usually no more than 10 minutes.

Employers should avoid the use of all substances with an MEL, and should find an alternative.

What is health surveillance?

Health surveillance is required under certain circumstances and requires the employers to regularly assess the health of their employees. If employees are exposed, for instance, to a substance that causes skin irritation, it may be necessary to check the condition of hands and arms by visual examination from time to time.

Health surveillance allows an employer the opportunity to monitor the effectiveness of the control measures in place.

If employees are exposed to fumes or dust, then routine lung tests or blood tests can be used.

Health surveillance can be carried out by a medical doctor or occupational nurse, or an employer can carry out simple assessments and refer to experts for advice.

What are the recommended steps when undertaking a COSHH assessment?

The HSE recommends an *eight-step* approach to a COSHH assessment as follows:

Step 1: Assess the risks
Step 2: Decide what precautions are needed
Step 3: Prevent or adequately control exposure
Step 4: Ensure that control measures are used and maintained
Step 5: Monitor exposure of employees (and others if appropriate)
Step 6: Carry out appropriate health surveillance
Step 7: Prepare plans and procedures to deal with accidents, incidents and emergencies
Step 8: Ensure employees are properly trained, informed and supervised

Regulation 6 of the COSHH Regulations 2002 requires employers to carry out a 'suitable and sufficient' assessment of the risks to health from using the hazardous substance.

'Suitable and sufficient' does not mean absolutely perfect, but the guidance that supports Regulation 6 lists a number of considerations that must be taken into account during the process.

The first step in providing a suitable and sufficient COSHH assessment is to ensure that the person carrying it out is *competent*.

Regulation 12(4) of the COSHH Regulations 2002 requires that any person who carries out duties on behalf of the employer has suitable information, instruction and training.

A competent person does not necessarily need to have qualifications as such but they should:

- have adequate knowledge, training, information and expertise to understand the terms 'hazard' and 'risk'
- know how the work activity uses or produces substances hazardous to health
- have the ability and authority to collate all the necessary relevant information
- have the knowledge, skills and experience to make the right decisions about risks and the precautions that are needed.

The person carrying out the assessment does not need to have detailed knowledge of the COSHH Regulations 2002 but needs to know who and where to go to obtain more information. They need to be able to recognise when they need more information and expertise. As the saying goes, 'a little knowledge is a dangerous thing'.

The COSHH assessment process can be broken down into eight steps as follows:

Step 1: Assess the risks
Identify the hazardous substances present on the site, or intended to be used on the site.

Consider the risks these substances present to your own employees and all others.

List all of the substances likely to be used. If the substances are on site, read the labels and look for the hazard warning symbols (Figure 5.1).

The symbols used on products will either be orange in colour with black symbols and writing, or will be the newer plain white diamond-shaped boxes with red edging and black symbols. The new symbols have been introduced to harmonise the use of hazard warning symbols across the world. Tables 5.1 and 5.2 show the old and new hazard warning symbols.

Obtain safety data sheets from the supplier or manufacturer.

A safety data sheet must be provided by the manufacturer or supplier and it must give information on, among other things, common usage, constituent ingredients, exposure limits, personal protective equipment, emergency procedures, first aid and spillage precautions.

If subcontractors are supplying the hazardous substances they must be required to submit manufacturers' data sheets together with their COSHH assessments.

The COSHH Regulations 2002 introduced a revision to the information needed for a COSHH assessment and now *all* COSHH assessments must have attached to them the product safety data sheet.

Figure 5.1 Hazardous substances symbols – older versions

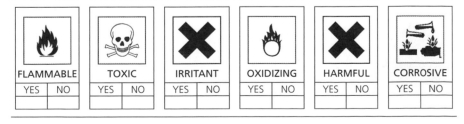

Table 5.1 Old and new hazard warning symbols in use on substances

Symbol	Description of risks
	Unstable explosives Explosives, divisions 1.1, 1.2, 1.3 and 1.4 Self-reactive substances and mixtures, types A and B Organic peroxides, types A and B
	Flammable gases, category 1 Flammable aerosols, categories 1 and 2 Flammable liquids, categories 1, 2 and 3 Flammable solids, categories 1 and 2 Self-reactive substances and mixtures, types B, C, D, E and F Pyrophoric liquids, category 1 Pyrophoric solids, category 1 Self-heating substances and mixtures, categories 1 and 2 Substances and mixtures which in contact with water emit flammable gases, categories 1, 2 and 3 Organic peroxides, types B, C, D, E and F
	Oxidising gases, category 1 Oxidising liquids, categories 1, 2 and 3
	Gases under pressure: – compressed gases – liquefied gases – refrigerated liquefied gases – dissolved gases
	Corrosive to metals, category 1 Skin corrosion, categories 1A, 1B and 1C Serious eye damage, category 1

Table 5.1 Continued

Symbol	Description of risks
	Acute toxicity (oral, dermal, inhalation), categories 1, 2 and 3
	Acute toxicity (oral, dermal, inhalation), category 4 Skin irritation, category 2 Eye irritation, category 2 Skin sensitisation, category 1 Specific target organ toxicity – single exposure, category 3
	Respiratory sensitisation, category 1 Germ cell mutagenicity, categories 1A, 1B and 2 Carcinogenicity, categories 1A, 1B and 2 Reproductive toxicity, categories 1A, 1B and 2 Specific target organ toxicity – single exposure, categories 1 and 2 Specific target organ toxicity – repeated exposure, categories 1 and 2 Aspiration hazard, category 1
	Hazardous to the aquatic environment – Acute hazard, category 1 – Chronic hazard, categories 1 and 2

Consider the risks of the hazardous substances identified to people's health. Remember, this applies to all people, not just employees. The COSHH Regulations 2002 clearly state that the effect of a hazardous substance on the health of 'other persons' must be clearly considered.

The assessment of the risk to health from using a hazardous substance is one of judgement. Sometimes it is not possible to know for sure what level of exposure will be harmful. In such cases, it is preferable always to err on the side of caution.

Table 5.2 Old hazard warning symbols that may be used on older chemicals

Symbol	Meaning	Description of risks
	Toxic (T) Very toxic (T+)	Toxic and harmful substances and preparations posing a danger to health, even in small amounts. If very small amounts have an effect on health the product is identified by the toxic symbol
	Harmful (Xn)	These products enter the organism through inhalation, ingestion or through the skin
	Highly flammable (F) Extremely flammable (F+)	(F) Highly flammable products ignite in the presence of a flame, a source of heat (e.g. a hot surface) or a spark. (F+) Extremely flammable products can be readily ignited by an energy source (flame, spark, etc.) even at temperatures below 0°C.
	Oxidising (O)	Combustion requires a combustible material, oxygen and a source of ignition; it is greatly accelerated in the presence of an oxidising product (a substance rich in oxygen)
	Corrosive (C)	Corrosive substances seriously damage living tissue and also attack the other materials. The reaction may be due to the presence of water or humidity
	Irritant (Xi)	Repeated contact with irritant products causes inflammation of the skin, mucous membranes, etc.
	Explosive (E)	An explosion is an extremely rapid combustion. It depends on the characteristics of the product, the temperature (source of heat), contact with other products (reaction), shocks or friction
	Dangerous for the environment (≪N)	Substances that are highly toxic to aquatic organisms, toxic to fauna or dangerous for the ozone layer

Some questions to consider are:

- How much of the substance is in use?
- How could people be exposed?
- Who could be exposed to the substance and how often?
- What type of exposure will they have?
- Could other people be exposed to the harmful substance?

Step 2: Decide what precautions are needed

The first responsibility for an employer is to eliminate the risk of using a harmful substance. This is an effective precaution to take but may not always be possible.

Next, consider whether an alternative substance could be used that is less hazardous than the original substance. If there is something on the market that does the job more safely, then it should be used.

If the substance cannot be eliminated or substituted, then it must be used with *suitable controls* so as to reduce any level of exposure to an acceptable limit.

Suitable controls may be:

- changing the way the work is done (e.g. painting instead of spraying)
- reducing concentrations of the substance
- modifying the work process (e.g. to reduce exposure time)
- reducing the number of employees and others exposed to the substance
- adopting maintenance controls and procedures
- reducing the quantity of the substance kept on site
- controlling the overall working environment (e.g. increasing ventilation, damping down dust, etc.)
- taking appropriate hygiene measures (e.g. easily accessible hand/arm washing stations to prevent skin absorption)
- enclosing the work activity.

Step 3: Prevent or adequately control exposure

Having decided what controls are required, the next step is to implement the controls on site.

The very nature of a construction site makes it quite difficult to implement some controls, but controls to consider are:

- substituting the substance
- excluding from the area workers not needed in the area

- introducing local exhaust ventilation
- changing work processes (e.g. reduce spraying in favour of brush application)
- creating zoned work areas
- undertaking tasks in the open air, where practicable.

Construction sites are particularly prone to a 'cocktail of substances', whereby different substances may combine to form a separate harmful substance. The principal contractor should coordinate the use of these different substances and complete the COSHH assessment for any 'communally' produced hazardous substance.

If there are no effective alternatives, it is permissible to issue operatives with personal protective equipment.

Personal protective equipment should be considered as the last resort – all other control measures must have been effectively considered and introduced if at all possible.

Personal protective equipment includes:

- face masks
- respiratory masks
- gloves and gauntlets
- goggles
- safety boots or shoes that are chemical resistant.

The control of harmful substances *must* be to a level that most workers can be exposed to day after day without adverse effects on their health.

Adequate control of harmful substances can be defined by referring to the occupational exposure limit (OEL) or workplace exposure limit (WEL) of a substance.

OELs and WELs are set at a value such that the substance is not likely to damage the health of workers who are exposed to it day after day.

If a substance has an OEL or WEL and the level of exposure is kept to the limit stated, then, as an employer, you will be deemed to have adequately controlled the risk.

Short-term, infrequent exposure to higher levels than the OEL or WEL are permissible but they must be the exception rather than the norm. Usually, such higher exposure will be because of an emergency (e.g. spillage, substance release or failure of exhaust ventilation).

Maximum exposure limits (MELs) are set for substances that may cause the most serious health effects, such as cancer and occupational asthma.

The use of any substance with an MEL must be reduced to a level that is below the MEL, and employers are required to ensure that exposure to such substances is *as low as is reasonably practicable.*

Step 4: Ensure that control measures are used and maintained

Employees and others must make proper use of the controls that an employer puts in place.

The person in control on a multi-occupied site must ensure that everyone, including contractors, follows their own COSHH assessments, and that others affected by the works also adhere to the controls identified as necessary to reduce exposure to harmful substances.

An employer has a duty to provide personal protective equipment (PPE) to employees, and to ensure that it is suitable and sufficient for their needs and is properly maintained.

Employees must report defects in their PPE to the employer or whoever has supplied the PPE (e.g. the principal contractor) or his representative.

All controls introduced onto the site to reduce the exposure to harmful substances must be adequately maintained so as to ensure that they are effective.

Engineering controls and local exhaust ventilation equipment must be regularly inspected, tested and examined, and records shall be kept. Local exhaust ventilation must be inspected every 14 months.

Records and test certificates must be kept for at least 5 years. If hazardous substances are used on a construction site the principal contractor should keep records and test certificates for any communal controls introduced to the site.

Step 5: Monitor exposure of employees (and others if appropriate)

If the risk assessment determines that there is a serious risk to health if people are exposed to a substance then health surveillance must be considered.

If any of the following apply, health surveillance will be essential:

- if control measures fail or deteriorate
- if the exposure limited could be exceeded
- if control measures are not maintained adequately.

Air monitoring must be carried out if there is a risk of exposure to the harmful substance due to inhalation.

Air monitoring is either site wide by environmental air monitoring or can be achieved by giving exposed workers personal air monitors.

Records of air monitoring should be kept for at least 5 years.

Step 6: Carry out appropriate health surveillance
Health surveillance is defined as:

> an assessment of the state of health of an employee, as related to exposure to substances hazardous to health, and includes biological monitoring.

Health surveillance will be necessary in the following circumstances:

- any exposure to lead fumes
- exposure to substances that cause industrial dermatitis
- exposure to substances that may cause asthma
- exposure to substances of recognised systemic toxicity (i.e. substances that can be breathed in, absorbed through the skin or swallowed and which affect parts of the body other than where they enter).

Any health surveillance carried out should be recorded and the records should be kept for at least 40 years. It is for the *employer* to decide whether health surveillance is necessary.

Step 7: Prepare plans and procedures to deal with accidents, incidents and emergencies
You need to plan and practice to cope with foreseeable accidents, incidents or emergencies. This means that you need:

- the right equipment to deal with an emergency (e.g. a spill), including protective equipment and decontamination products
- the right procedures to deal with a casualty
- the right people trained to take action
- the right arrangements to deal with the waste created.

Think about how you would make such information available to the emergency services. Everybody needs to know your emergency plans.

Involve safety representatives and employees.

Step 8: Information, instruction and training

All employees expected to use or to come into contact with substances hazardous to health shall receive suitable information, instruction and training.

The employer should ensure that all employees and contractors have adequate records for training their operatives in the health and safety risks of using hazardous substances.

The employer should ensure that the site induction training covers the use of hazardous substances on the site and the control measures to be followed.

All operatives, visitors, contractors, etc. are entitled to see the COSHH assessment and the attached safety data sheet.

Information may be by way of safety notices, toolbox talks, site rules, etc.

Instruction generally covers one-to-one exchanges on how to do something, what to use, what personal protective equipment to wear, etc. The health effects of using the hazardous substance should be clearly discussed.

It is useful to keep records of any exchange of information on COSHH in the site records book.

What are an employer's responsibilities regarding information, instruction and training for employees?

The employer should inform and train all employees on the hazards and risks from substances with which they work, and on the use of control measures developed to minimise the risks.

For control measures to be effective, people need to know how to use them properly. Most importantly, people need to know why they should be bothered to work in a certain way and use the controls as specified; they need to be motivated.

Motivation comes from understanding what the health risks are and, therefore, why the control measures are important. It also comes from users having confidence in the control measures and believing that they will protect their health. If the health risk is serious (e.g. silicosis, cancer, asthma, allergic dermatitis or a blood-borne disease such as HIV) and is chronic or latent in nature, a good appreciation of the risk is especially important. With latent or delayed risks, exposure can often be excessive, with no short-term warning, such as an odour or irritation, to indicate that anything is amiss. People who will potentially be exposed need to be told, clearly and honestly, why they should use the control measures, and the consequences in terms of ill health, if they do not use them.

People need to know how control measures work in order to use them correctly and to recognise when they are not working properly. This means training operators who are directly involved, and also supervisors and managers. This is so that everyone can identify when controls are being used in ways that reduce their effectiveness. It is important to know whether the individual is working in a way that reduces the effectiveness of control measures because:

- there is no other way of doing the job
- he or she does not know any better.

If the control measures are difficult to use or get in the way of doing the job, they will need redesigning. If the control measures are well designed and tested but are still misused, then the individual needs retraining and motivating.

Most control measures involve methods of working, which means that, at the design stage, it is essential to ask workers and supervisors for their views on how best to do the work, so that exposure is minimised. They should be asked whether a proposed method of working is practical and how to get the best out of the proposed control measures. People who are actively involved in the development of control measures are more likely to 'own' them and respond positively to the new ways of working that may be required. Easily followed, convenient and simple procedures that minimise exposure and are built in to the working method are more likely to be followed.

Is there a relatively easy way to remember the steps in a COSHH assessment?

Yes! A very useful acronym is the word SITE (Substance—Individual—Task—Environment). Remember that risk assessments should be site specific and COSHH assessments substance specific, so SITE is quite useful.

Examples of questions under each heading are listed below. There may be a few more that would be relevant, depending on the circumstances.

Substance

- How will it get into the body?
- What harm could it do to the body?
- Are there any workplace exposure limits?
- What form is the substance in (solid, liquid or airborne)? Would this lead to significant exposure?

- What is the chance of exposure? This may be affected by the training and information people have received and how reliable and suitable the control measures are.
- How often may people be exposed?
- Could the substance be mixed (intentionally or accidentally) with another one, which could be dangerous?
- Might the substance produce dangerous fumes or be a fire risk?

Individual
- Who will be using and/or be exposed to the substance?
- Are there any people who would be particularly vulnerable (e.g. respiratory issues or pregnancy)?
- Are users capable of using the substance safely?

Task
- How will the substance be used or produced?
- Will a substance be used in line with the manufacturer's instructions?
- What other hazards may be involved (e.g. working at height)?
- What will the user do next (e.g. working up a ladder or driving after using a high-solvent paint might be dangerous)?

Environment
- Where is the substance going to be used and produced, and who might this affect?
- Is there anything in the environment that could be incompatible with the substance (e.g. a naked flame in the presence of flammable substances, or other substances that could react chemically with the substance)?

Remember that all these factors are linked, so they should never be considered in isolation.

How might the risk of accidents from hazardous substances be reduced?
- Check that packages and containers are in good condition, so as to avoid leaks. Make sure that gases, fumes, vapours and dusts are extracted at their point of origin. Wear a respirator if necessary. Watch out for possible sources of fire.
- Keep dangerous products only in appropriate containers, property labelled. Never transfer them into bottles such as lemonade or beer bottles, or other food containers. This type of practice causes serious accidents every year. Dangerous products should preferably be kept locked away when not in use.
- Avoid contact with the mouth. Do not eat, drink or smoke when using dangerous substances or when in a place where they are used.

Example of an approach to COSHH

A retail garden centre with a small facility for manufacturing garden furniture
Full-time workers could be exposed to the following:

- spillage during handling of garden chemicals
- wood dust, paints, solvents, preservatives and adhesives in the furniture workshop
- diesel fumes from the forklift in storage buildings
- dipping bulbs in organophosphorous pesticide.

Ancillary workers could be exposed to the following:

- cleaners mopping up spillages
- cleaners using hazardous cleaning materials (e.g. bleach)
- maintenance workers repairing contaminated racking.

Contractors and visitors (including the public) could be exposed to a variety of hazards, especially where there is little supervision or control of access to various parts of the premises.

Supervisors and managers are likely to be exposed to all the circumstances listed, although their exposure will usually be of greatly reduced length and intensity.

Students and other casual workers might be employed to stack shelves. There is some scope for them to come into contact with spillages of hazardous substances during this work. They may also find themselves helping out in other areas when there is a temporary need for an 'extra hand'.

Office workers will be exposed to the normal range of substances found in offices. They may also have incidental contact with other work activities on the site. For instance, do they regularly have to pass through the woodworking shop to get to the office? There will usually be little scope for effects on people outside the premises, but instances where there is might include:

- spray drift from pesticide treatment of outside plants
- disposal of spent bulb-treatment solution
- fumigation of on-site production (as opposed to display) glasshouses.

(Extract from the HSE website)

What is REACH?

Registration, Evaluation, Authorisation and Restriction of Chemicals (REACH) is a European Union (EU) regulation (EU 1907/2006) concerning the registration, evaluation, authorisation and restriction of chemicals. It came into force on 1 June 2007 and replaced a number of European directives and regulations with a single system.

Aims

REACH has several aims:

- to provide a high level of protection of human health and the environment from the use of chemicals
- to make the people who place chemicals on the market (manufacturers) and importers responsible for understanding and managing the risks associated with their use
- to allow the free movement of substances on the EU market
- to enhance innovation in and the competitiveness of the EU chemicals industry
- to promote the use of alternative methods for the assessment of the hazardous properties of substances (e.g. quantitative structure–activity relationships and read across).

Scope and exemptions

REACH applies to substances manufactured in or imported into the EU in quantities of 1 tonne or more per year. Generally, it applies to all individual chemical substances on their own, in preparations or in articles (if the substance is intended to be released during normal and reasonably foreseeable conditions of use from an article).

Some substances are specifically excluded:

- radioactive substances
- substances under customs supervision
- the transport of substances
- non-isolated intermediates
- waste
- some naturally occurring low-hazard substances.

Some substances, covered by more specific legislation, have tailored provisions, including:

- human and veterinary medicines
- food and foodstuff additives
- plant-protection products and biocides.

Other substances have tailored provisions within the REACH legislation, as long they are used in specified conditions:

- isolated intermediates
- substances used for research and development.

How does an employer carry out a risk assessment under the Dangerous Substances and Explosive Atmospheres Regulations 2002?

The Dangerous Substances and Explosive Atmospheres Regulations 2002 (DSEAR) are concerned with protection against risks from fire, explosion and similar events arising from dangerous substances used or present in the workplace. From June 2015 the DSEAR also cover gases under pressure and substances that are corrosive to metals.

Examples of substances coming under DSEAR controls are:

- petroleum
- flammable gases (e.g. liquid petroleum gas)
- gases under pressure
- solvents, paints, varnishes and flammable gases
- dusts from machining and sanding operations
- dusts from foodstuffs
- substances corrosive to metal.

The DSEAR require employers (or self-employed persons) to:

- carry out a risk assessment before commencing any new work that involves a dangerous substance
- if there are five or more employees, record the findings in writing as soon as is practicable after the assessment
- take steps to eliminate the risks
- ensure that the workplace and work equipment are safe during operation and maintenance
- detail any hazardous zones
- detail any special measures of cooperation between more than one employer
- introduce arrangements for dealing with accidents and emergencies.

The risk assessment involves the identification and careful assessment and examination of the dangerous substances present in the workplace, the work practices and activities using those substances, and what the consequences will be if something goes wrong.

The risk assessment must consider the risks not only to employees but also to members of the public.

Employers need to establish what they need to do to reduce or eliminate risks as far as is reasonably practicable so as to ensure everyone's safety when using dangerous substances.

The risk assessment *must* be completed before work commences with a dangerous substance, and the control measures identified as being necessary must be implemented *before* works commence.

The risk assessment process should be the same as for all risk assessments:

- identify the hazard
- identify who may be harmed and how
- evaluate the risks and decide on the control measures necessary
- record the findings
- audit and review the risk assessment.

What are some control measures that could be implemented?

The following list gives some examples of control measures but the risk assessment should help identify what controls are needed in the actual environment in which the dangerous substance is present:

- reduce the quantity of dangerous substance to a minimum
- avoid or minimise releases
- control releases at source
- prevent the formation of an explosive atmosphere
- collect, contain and remove any releases to a safe place
- avoid ignition sources
- avoid adverse conditions that could lead to danger (e.g. avoid excessive temperatures)
- keep incompatible substances apart.

Other steps could be taken that will mitigate the risks associated with a dangerous substance:

- reduce the number of employees exposed
- provide plant that is explosion proof
- provide explosion suppression or explosion relief equipment
- take measures to minimise or control the spread of fire or explosion
- provide suitable personal protective equipment
- design and construct the workplace to reduce the risks from dangerous substances
- choose appropriate equipment and work systems that reduce risks
- implement safe systems of work (e.g. permit to work procedures).

How will an employer identify a dangerous substance?

The employer will need to carry out two steps:

- check whether the substance has been classified as
  - explosive
  - oxidising
  - extremely flammable
  - highly flammable
  - flammable
- assess the physical and chemical properties of the substance and the circumstances of the work involving the substance, to see what will create a safety risk to persons.

Any substance labelled

- explosive
- oxidising
- flammable (in all categories)

is a dangerous substance, and the DSEAR apply.

COSHH checklist

1.	Have you a complete inventory of substances used/generated in the workplace?	Yes	No
2.	Have you identified any substances hazardous to health?	Yes	No
3.	Have you gathered information about the substances, the work and working processes, i.e.:		
	■ What hazards are involved?	Yes	No
	■ Who could be exposed and how?	Yes	No
4.	Have you evaluated the risks to health (either on an individual or group basis)?		
	■ The chance of exposure occurring	Yes	No
	■ The level of exposure that could happen	Yes	No
	■ The duration of the exposure	Yes	No
	■ The frequency of the exposure	Yes	No
5.	Have you decided what needs to be done in terms of:		
	■ Preventing or controlling exposure?	Yes	No
	■ Maintaining control measures?	Yes	No
	■ Using control measures?	Yes	No
	■ Any monitoring or surveillance?	Yes	No
	■ Information, instruction and training?	Yes	No
6.	Have you decided to record the assessment?	Yes	No
7.	If 'yes' to (6), have you decided on the extent, presentation and format of record?	Yes	No
8.	Have you decided when each assessment should be reviewed?	Yes	No
9.	Have you established a system or procedure to manage and record the above elements?	Yes	No

CONTROL OF SUBSTANCES HAZARDOUS TO HEALTH ASSESSMENT SHEET

Company: ....

Address: ....

Contact: ....

Product: ....

Job task: ....

Application: ....

Equipment:

CONTROL MEASURES

For users: ....

For persons in location: ....

EMERGENCY PROCEDURE

Emergency contact No.: ....

Spillage arrangements: ....

Consumption arrangements: ....

Contact arrangements: ....

PRODUCT

SAFETY DATA SHEET ATTACHED: YES/NO

RISK IDENTIFICATION

Hazardous component(s): ....

Hazardous nature of component(s): ....

Health hazards (known): ....

Persons affected: ....

Duration of exposure: ....

Level of exposure: ....

Risk category: High ☐   Medium   ☐   Low ☐

OCCUPATIONAL EXPOSURE LIMITS

Occupational exposure standard/workplace exposure limit: . . . .

Maximum exposure limit (8 hour TWA): . . . .

Maximum exposure limit (15 minute TWA): . . . .

Control measures of exposure: . . . .

Monitoring requirements: . . . .

OTHER CONTROL MEASURES

Training: . . . .

Health surveillance: . . . .

Reassessment: . . . .

Name of person undertaking assessment: . . . .

Date of assessment/revision: . . . .

Risk Assessments: Questions and Answers
ISBN 978-0-7277-6076-0

ICE Publishing: All rights reserved
http://dx.doi.org/10.1680/raqa.60760.089

Chapter 6
Manual handling

What are employers responsible for in respect of manual handling at work?

The Manual Handling Operations Regulations 1992 apply to all manual handling activities carried out by employees while at work.

Employers must, as far as is reasonably practicable, avoid the need for employees to undertake any manual handling activities while at work that involve risk of injury.

Despite the above requirement, manual handling is a common cause of work-related injury. In some cases, poor manual handling can lead to permanent disability and physical impairment.

Employers must undertake a risk assessment of all manual handling activities and determine a hierarchy of risk control in order to minimise risks of injury and ill health.

Information of the weight of objects to be manually handled must be given to employees. This can be general information or more specific and precise product information. Many manufacturers and suppliers of products and equipment display the weight of the item on packaging, delivery notes, etc.

What are the costs of poor manual handling to both businesses and society?

Injuries sustained by employees while they are handling, lifting or carrying items at work account for nine out of ten of all over 7 day injuries.

Over 1 million people are reported to have suffered 'illness' from musculoskeletal disorders, and the prevalence rate is increasing compared with the early 1990s.

Statistics from the Health and Safety Executive for 2013/2014 show that approximately 900 000 working days were lost to manual handling injuries.

The National Health Service (NHS) has one of the highest incidence rates for musculoskeletal injury. Approximately 50% (i.e. approximately 5000) of all reported 'over 7-day injuries' were due to handling, lifting and carrying.

Back injury is not the only type of injury sustained from manual handling. Injuries can affect:

- hands
- feet
- arms
- legs

as well as provoke lacerations and fractures.

Many manual handling injuries are the result of poor practices being followed over lengthy periods of time and not a 'one-off' manual handling activity.

On average, each injury takes 20 days for recovery, and in some instances disability is permanent. Costs to business will be huge and are often hidden in real terms. Costs to be considered are:

- sick pay
- loss of a skilled employee
- replacement/temporary staff
- reduced productivity
- investigation time
- civil claims
- criminal prosecution
- increased insurance premiums.

Back injuries represent the biggest single group of claims for incapacity benefit.

Costs for manual handling injuries have been estimated at £6 billion in lost production.

Case study

In 2002, an ambulance worker received compensation of £140 000 in an out-of-court settlement with his employer for serious back injuries sustained during the course of his employment.

The employee was lifting a patient when two wheels came off the stretcher he was carrying. He then had to bear the patient's weight for 5 minutes. As a result he damaged his lower back and right leg.

Damages were claimed against the NHS Trust because the stretcher had been modified to fit into the ambulance and was not fit for purpose. The Trust admitted liability.

Consideration is being given by the NHS Trust to instigating legal action against the stretcher manufacturer.

What steps should be taken in respect of carrying out a risk assessment for manual handling?

In the first instance, it would be sensible to conduct a general assessment to see if manual handling activities give rise to hazard and risk, as not *all* manual handling will.

Remember that manual handling includes:

- lifting
- pushing
- pulling
- shoving
- lowering
- carrying.

Consider the size and shape of the load and the best way to handle it. If the load is difficult or heavy, seek assistance.

Consider also where the load is going. Is the pathway clear and free from obstruction? Is the place where the load is to go ready to receive it?

Can lifting devices be used, or can the load be split to make carrying easier?

Manual handling involves pushing and pulling as well as lifting. Can any of these jobs be mechanised?

Complete a manual handling risk assessment.

What needs to be considered in a detailed manual handling risk assessment?

The tasks

Do the tasks involve:

- holding loads away from the body
- twisting, reaching or stooping
- strenuous pushing or pulling
- unpredictable movement of loads
- large vertical movement
- long carrying distances
- repetitive handling
- insufficient rest time
- a work rate imposed by a process?

The loads
Are the loads:

- heavy, bulky or unwieldy
- difficult to hold
- unstable or unpredictable
- intrinsically harmful (e.g. sharp)?

The working environment
In the working environment, are there:

- constraints on posture
- variations in level
- poor floors
- hot, cold or humid conditions
- strong air movements
- poor lighting
- restrictions on movement or posture from clothes or personal protective equipment?

Individual capacity
Does the job:

- require unusual capability
- endanger those with a health problem
- endanger pregnant women
- require special information or training?

Case study

Types of manual handling in licensed premises
The following activities undertaken routinely in most pubs are likely to present particular risks in terms of manual handling operations:

- the delivery and removal of full and empty kegs, boxes, barrels, crates and gas cylinders
- stacking of full kegs and barrels
- moving kegs, barrels, etc. within the cellar or storeroom
- shifting of casks
- moving loads between floors – carrying crates from the cellar to the bar
- lifting buckets of water or pipe-cleaning containers
- lifting gas cylinders

- putting items on shelves and taking items off shelves
- moving furniture, equipment, etc.
- food deliveries
- the removal of glass bottle skips
- carrying empty glass baskets and crates
- changing optics
- lifting glasswasher trays
- carrying tills or money drawers
- carrying money or change
- moving AWP (amusement with prize) machines
- assisting with entertainment equipment.

What are some of the ways of reducing the risks of injury from manual handling?

The tasks

Can you:

- reduce the amount of twisting and stooping
- avoid lifting from floor level or above shoulder height
- avoid strenuous pushing or pulling
- reduce carrying distances
- avoid repetitive handling
- vary work, allowing one set of muscles to rest while another is used?

The loads

Can you make the loads:

- lighter or less bulky
- easier to hold
- more stable
- less damaging to hold

and

- have you asked your suppliers to help in this regard?

The working environment

Can you:

- improve the workplace layout to improve efficiency
- remove obstructions to free movement

- provide better flooring
- avoid steps and steep ramps
- prevent extremes of hot and cold
- improve lighting
- consider less restrictive clothing or personal protective equipment?

Individual capacity
Can you:

- take better care of those with physical weaknesses or who are pregnant
- give your employees more information (e.g. about the range of tasks they are likely to face)
- provide training?

What are good handling techniques?
The following important points to bear in mind when handling a load are illustrated using a basic lifting operation as an example.

Planning
Plan the lift. Where is the load to be placed? Use appropriate handling aids if possible. Do you need help with the load? Remove obstructions. For a long lift, such as from the floor to shoulder height, consider resting the load midway on a table or bench to change grip.

Positioning feet
Feet should be placed apart, giving a balanced and stable base for lifting (tight skirts and unsuitable footwear make this difficult). The leading leg should be as far forward as is comfortable and, if possible, should be pointing in the direction you intend to go.

Good posture
When lifting from a low level, bend with the knees, but do not kneel or overflex the knees. Keep the back straight, maintaining its natural curve (tucking in the chin helps). Lean forward a little over the load if necessary to get a good grip. Keep the shoulders level and facing in the same direction as the hips.

Holding the load
Try to keep the arms within the boundary formed by the legs. The best position and type of grip depends on the circumstances and individual preference, but must be secure. A hook grip is less tiring than keeping the fingers straight. If you need to vary the grip as the lift proceeds, do it as smoothly as possible.

Lifting

Keep the load close to the trunk for as long as possible. Keep the heaviest side of the load next to your trunk. If a close approach to the load is not possible, slide it towards you before you try to lift. Lift smoothly, raising the chin as the lift begins, keeping control of the load.

Movement

Move the feet instead of twisting the trunk when turning to the side.

Adjustment

If precise positioning of the load is necessary, put it down first, then slide it into the desired position.

Risk assessment: handling a 20-litre drum

The task
Between twenty and thirty 20-litre drums were moved from the delivery area to the storage area. They were then moved from the storage area to the point of use.

The problem
To move the 20-litre drums to the storage area, maintenance staff had to carry them down 20 steps. No injuries had been reported, but the task was identified as having a high potential risk of causing musculoskeletal injuries.

The solution
The storage area was reorganised and moved to be near the delivery area so that the need to carry the drums down the steps was eliminated.

Two sack barrows were provided so that the drums could be easily moved on the same level to the areas where they are needed.

The benefits
- Easier manual handling.
- More efficient use of time – less double handling.
- Reduction in potential manual handling injuries.

What are the guideline weights for lifting or manual handling?

	Women	Men
Shoulder height		
Arms extended	3 kg	5 kg
Near to body	7 kg	10 kg

Elbow height

Arms extended	7 kg	10 kg
Near to body	13 kg	20 kg

Thigh height

Away from body	10 kg	15 kg
Near to body	16 kg	25 kg

Knee height

Away from body	7 kg	10 kg
Near to body	13 kg	20 kg

Lower leg height

Away from body	3 kg	5 kg
Near to body	7 kg	10 kg

Each category above indicates the guideline weights for lifting and lowering loads.

Heavier weights can be handled more safely if they are held close to the body. Carrying objects at arm's length creates extra strain on the spine and muscles, and therefore lower weights are recommended.

For the weights listed it is assumed that the load is readily grasped with both hands and that the operation takes place in reasonable working conditions with the lifter in a stable body position.

Any operation involving more than twice the guideline weights should be rigorously assessed, even for fit, well-trained individuals working under favourable conditions.

Twisting
Reduce the guideline weights if the lifter twists to the side during the operation. As a rough guide, reduce the weights by 10% if the handler twists beyond 45°, and by 20% if the handler twists beyond 90°.

Frequent lifting and lowering
The guideline weights are for infrequent operations – up to about 30 operations per hour – where the pace of work is not forced, adequate pauses to rest or use different muscles are possible, and the load is not supported for any length of time. Reduce the weights if the operation is repeated more often. As a rough guide, reduce the weights by 30% if the operation is repeated five to eight times a minute, and by 80% if the operation is repeated more than 12 times a minute.

Exceeding the guidelines

The risk assessment guidelines are not safe limits for lifting. But work outside the guidelines is likely to increase the risk of injury, so you should examine it closely for possible improvements. You should remember that you must make the work less demanding if it is reasonably practicable to do so.

Has 'ergonomics' anything to do with manual handling?

Yes and no. Ergonomics is the science concerned with the 'fit' between people and their work and surroundings.

Ergonomics aims to make sure that tasks, equipment, information and the environment suit each worker. So it could include manual handling activities, but it is more likely to consider *how* the job is done rather than *what* is lifted.

How can ergonomics improve health and safety?

Applying ergonomic principles to the workplace will:

- reduce the potential for accidents
- reduce the potential for injury and ill health
- improve performance and productivity.

Equipment, controls, operating panels, isolation switches, etc. should all be designed for ease of use, but how many times are switches awkward to get at, requiring twisting and contortion to use them?

A machine with a control panel that the operator is required to use could be the cause of accidents and injury if:

- the switches and buttons could be easily knocked on or off, thereby starting or stopping the machine by mistake
- the warning lights or switches are unusual colours or the reverse colours to what is usually expected (e.g. red for 'go', green for 'danger'); also, colours may be important as many people have red/green colour blindness
- the instruction panel and information given on how to use the controls is complicated or too detailed, causing operator confusion and inappropriate actions.

Ergonomics would look at all of the above issues and 'design out' the hazards associated with the control panel on the machine. The location of controls would be considered in order to cut down on 'repetitive strain' injuries.

What kind of manual handling problems can ergonomics solve?

The application of ergonomics is typically known as a means of solving physical problems, and in respect of manual handling these would be:

- loads that are too heavy or bulky
- loads that need to be lifted from the floor or above shoulder height
- repeated repetitive lifting
- tasks that involve awkward postures, twisting or bending
- loads that cannot be gripped properly
- tasks that need to be carried out in poor environmental conditions (i.e. wet floors, poor lighting, cramped space or restricted head height)
- tasks that are carried out under too great a time pressure and without adequate rest periods.

Any of the above situations can lead to operator tiredness and exhaustion, which increases the risk of accidents and injury.

Ergonomics is about finding solutions for alternative ways of doing the job.

Is it appropriate to have a 'no lifting' policy within the work environment?

'No lifting' policies seem to be common in the care services industry where there is a lot of manual handling of people and equipment.

Such an all-encompassing policy may be workable in some organisations but in reality it may be extremely difficult to enforce such a policy.

The Manual Handling Operations Regulations 1992 require that 'hazardous' lifting or manual handling is eliminated or reduced to an acceptable level. Manual handling in all its forms cannot be eliminated but controls can be put in place to reduce the likelihood of injury.

A 'no lifting' policy would require mechanical aids to be provided to assist with lifting. In some circumstances the use of a lifting device may create a greater hazard than the lifting itself.

Rather than a 'no lifting' policy it would be appropriate to have a 'lifting' policy that sets out what type of manual handling is undertaken, what the hazards and risks are, and how control measures can be used to reduce the risks.

Case study

Risk assessment format for care services

Consider activities to be undertaken during the day and also at night.

A risk assessment for use in the care services (e.g. for a care home) should be set out in a simple format so that it is possible to quickly assimilate what equipment, techniques and numbers of staff are appropriate.

The following should be included:

- details of the individual, including their height and weight
- the extent of the individual's ability to support his or her own weight, and other relevant factors (e.g. pain, disability and tendency to fall)
- any problems the individual has with comprehension or cooperational behaviour
- recommended methods for relevant tasks such as sitting, visiting the toilet, bathing, transfers and movement in bed
- the minimum number of staff needed to help
- the need for and availability of lifting or moving equipment
- other relevant risk factors
- what training and individual capability of the care worker is required.

What are some of the key solutions to manual handling problems?

There are several ways in which manual handling problems can be reduced or eliminated.

These include:

- avoiding manual handling through automation or changing the overall process
- redesigning the load
- redesigning the task
- redesigning the working environment
- introducing mechanical handling aids.

The changes do not have to be expensive or complicated to be effective. Simple solutions are often better.

Principles for developing successful solutions to manual handling problems

- Prioritise your activities.
- Tackle serious risks affecting a number of employees before a tackling an isolated complaint of minor discomfort.
- Find and evaluate a few possible solutions.
- Try out ideas on a small scale, and if necessary modify them prior to full implementation.
- Monitor the solutions to make sure they remain effective.
- Keep abreast of new technologies.

MANUAL HANDLING – RISK ASSESSMENT FORM

Job description: . . . .

Who is undertaking the tasks? . . . .

What are the hazards involved in the job? . . . .

. . . .

What are the risks? . . . .

. . . .

How likely are the risks? . . . .

. . . .

What control measures are needed to reduce or eliminate the risks?

Those currently in place: . . . .

. . . .

Those which need to be implemented: . . . .

. . . .

When should control measures be implemented? . . . .

. . . .

When should the risk assessment be reviewed? . . . .

. . . .

Date assessment completed: . . . .

By whom: . . . .

Case studies

Manual handling prosecution for lack of training and assessment

The importance of conducting manual handling risk assessments and implementing safe systems of work with suitable manual handling training is demonstrated by this case from October 2010.

A Warwickshire bathroom company was prosecuted after a worker suffered crush injuries when unloading items from a delivery vehicle. Two workers were involved – one passing boxes from the vehicle down to the other, who was standing on the ground. The boxes weighed 25–50 kg each and were sat on a shrink-wrapped pallet on the vehicle. The driver cut away the shrink wrap on one pallet and several items fell out onto the worker below. The injured worker was taken to hospital and had a month off work with damaged ligaments in his wrist.

The company was found guilty of failing to conduct a risk assessment, failing to provide training and failing to use a safe system of work. It was fined £4000 and ordered to pay costs of £2500.

The company has subsequently introduced a safe system of work using a vehicle with a tail lift and ensuring that all palleted goods are moved by a fork-lift truck. In addition, all staff have received training in manual handling.

Lack of manual handling training causes resident's death

A South Lanarkshire care home provider was fined £57 000 for failing to adequately train employees at its Uddingston facility, which led to the death of an 88-year-old resident.

A new employee at the care home had neither been trained nor adequately supervised in moving and handling the resident, who required two employees for assistance when washing, dressing and undressing. The new employee, working alone, was helping the resident transfer from her shower chair to her bed when the resident slipped and broke her neck. Although a risk assessment had determined the resident required the assistance of two people, the resident's information was not updated and the new employee was unaware of the requirement.

Risk Assessments: Questions and Answers
ISBN 978-0-7277-6076-0

ICE Publishing: All rights reserved
http://dx.doi.org/10.1680/raqa.60760.103

Chapter 7
Noise at work

What is noise at work?

Noise at work comes in many different forms, including from machinery, music and factory processes. Noise at work can damage hearing and in some cases lead to deafness, depending on how loud the noise is and for how long a person is exposed to it. All employers have a duty under health and safety law to reduce the risk of hearing damage to their employees by controlling exposure to noise.

Noise can also be a safety hazard at work by interfering with communication and warning sirens, making them harder to hear.

How is noise measured?

Noise is measured in decibels, usually abbreviated to dB. Sometimes you may see it written as dB(A), this is known as an A weighting and is an average of the noise level. The dB is a logarithmic unit used to describe a ratio. Due to the logarithmic effect of the dB scale a small increase in dB, such as 3 dB, can actually mean that the level of noise has doubled, so what might seem like a small increase in noise level is in fact a very significant one.

What do the Control of Noise at Work Regulations 2005 require?

The regulations place a duty on all employers whose employees are or may be exposed to noise while at work. The legislation sets out the action levels listed in Table 7.1 and states what action should be taken at each level by the employer to control noise levels.

The action levels are the noise exposure levels at which employers are required to take certain steps to reduce the harmful effects of noise on their employees. The action levels shown in the table are a daily or weekly average of noise exposure.

Do the Control of Noise at Work Regulations 2005 apply to all workplaces?

The regulations, which came into force in April 2006, apply to all premises where noise may affect a person at work, whether it is a bar person working in a nightclub or a machinist working on a factory production line.

Table 7.1 Action levels and actions required by the Control of Noise at Work Regulations 2005

Exposure action level	dB	Action required if level is exceeded
Lower exposure action level	80 dB(A)	Carry out a risk assessment Make suitable ear protection available Implement a maintenance programme for ear protection Must implement a training programme
Upper exposure action level	85 dB(A)	Reduce the noise at source Implement ear protection zones Ear protection must be provided and must be used by employees (the use of hearing protection is mandatory if the noise cannot be controlled by any other measure) Health surveillance provided for employees

The regulations set noise levels at which action must be taken to control the noise at work (see Table 7.1).

How do you determine if you have a noise problem at work?

The factors that need to be taken into consideration when determining if noise is a problem at work are:

■ how loud the noise is
■ how long people are exposed to it.

As a simple guide you will need to do something about the noise if any of the following apply:

■ the noise interferes with the day-to-day work activities for most of the day (e.g. a busy street or a vacuum cleaner being used non-stop)
■ employees have to raise their voices to carry out a normal conversation
■ noisy tools or machinery are used for more than half an hour throughout the working day
■ employees work in a noisy industry, such as construction, road repair, engineering, canning, production, manufacture, foundry, paper or board making
■ there is noise in the workplace due to machinery impacts, such as hammering, pressing, forging, pneumatic equipment or explosive sources.

What are some typical noise levels associated with construction?

Sound pressure in weighted decibels: dB(A)	Situation
140	*Peak action level, immediate irreversible damage* Jet at 30 m
130	*Threshold of pain* Pneumatic breaker (unsilenced) at 1 m
120	Pneumatic digger 600 hp, scraper at 2 m
110	Rock drill, diesel hammer driving sheet steel at 10 m
100	Scrabbling 7 hp road roller on concrete at 10 m
95	Concrete pouring
85	*Second action level (Control of Noise at Work Regulations 2005)* Drilling/grinding concrete
80	*First action level (Control of Noise at Work Regulations 2005)* Scaffold dismantling at 10 m, 8 hp diesel hoist at 10 m
70	5 hp power float at 7 m
60	Typical office
50	Living room

What to do if you have a noise problem?

If any of the above apply then an assessment of the risks will need to be carried out to decide whether further action is needed. This is known as a risk assessment. The aim of the risk assessment is to provide you with information so that a decision can be made about what needs to be done to ensure the health and safety of employees who are exposed to noise. In some cases measurements of noise may not be necessary: it is about collecting as much information as possible.

The noise risk assessment should contain the following information:

- who is at risk from noise
- who may be affected
- an estimate of the employee's exposure to noise, compared with the lower and upper exposure action levels

- what needs to be done to comply with the law – this is often called 'noise control measures', and may include the provision of hearing protection, such as ear defenders or earplugs; if noise control measures are required then the type should be included in the risk assessment
- details of any employees who need to be provided with health surveillance, and whether any particular employees are at risk because of the nature of their work.

There is no right or wrong way to complete a risk assessment. The law requires that it is 'suitable and sufficient'.

A risk assessment must contain suitable information to be useful to employees such that they understand what hazards they may be exposed to when carrying out the task.

How do I estimate an employee's exposure to noise?

The key is to ensure that the estimated employees' exposure is a true reflection of the work that they do, and should therefore take into account the following:

- the work that they do and may be doing in the future
- the way in which they do the work
- how the work might vary from one day to the next.

The best way to collect this information is to speak to the employees concerned, as the estimate must be based on correct information. The suppliers of machinery will also be able to provide data sheets that show the noise level associated with the use of a particular piece of machinery.

The information gathered must be recorded in a risk assessment. The risk assessment should set out what you have done and what you are going to do to ensure that employees' exposure to noise is controlled. There should also be a timetable showing when the measures will be implemented, along with who will be responsible for the work.

Once a risk assessment has been carried out, the information should be used to determine if a noise assessment is required.

In order for employers to be able to ensure that they are controlling their employees' exposure to loud noise they need to know which employees are at risk, and what that level of risk is.

Employers whose employees are exposed to noise while at work may have to implement measures to control their employees' exposure to that noise, if the level of noise is deemed to exceed the action levels stated in the regulations.

The best way of doing this is to have a noise assessment carried out. The Control of Noise at Work Regulations 2005 require the employer to carry out an adequate noise assessment, which will help to provide the information required to control noise in the workplace. In addition, the noise assessment assists in determining the most suitable hearing protection to provide for employees, whether ear protection zones are required and, if so, where they should be. The noise assessment document helps employers in their compliance with duties relating to controlling noise exposure.

What is the difference between a noise risk assessment and noise assessment?

A noise risk assessment is the first step in the process of determining whether employees are exposed to noise at work. The risk assessment process should be used to gather as much information as possible about the type of work that is carried out, who is at risk from the work, what the estimated level of exposure is, how employees may be affected, and what is or will be done to reduce the noise that employees may be exposed to.

If the risk assessment suggests that you have a noise problem, a competent person may need to be employed to measure the noise and determine the representative daily or weekly personal noise exposure; this is called a 'noise assessment'. The noise assessment includes measuring the sound pressure level at the different places where the employee works and for the different tasks carried out during the day. The average is calculated from these values and the time spent in each place or on each task. Information on getting started with a noise risk assessment is given in the Health and Safety Executive (HSE) publication *Noise at Work: A Brief Guide to Controlling the Risks* (INDG362, 2012) and more detail can be found in the HSE publication *Controlling Noise at Work* (L108, 2005).

Is it a legal requirement to carry out a noise assessment?

In order for employers to be able to ensure that they are controlling their employees' exposure to loud noise they need to know which employees are at risk, and what that level of risk is.

Employers whose employees are exposed to noise while at work may have to implement measures to control their employees' exposure to that noise if the level of noise is deemed to exceed the action levels stated in the regulations.

The best way of doing this is to carry out a noise assessment.

Is there a standard format for a noise assessment?

No, although there is certain information that should be contained in the noise assessment.

The HSE has published guidance on noise assessments, which provides a guide as to what the noise assessment should contain and a checklist of requirements to ensure that the assessment complies with the requirements of the Control of Noise at Work Regulations 2005. The guidance is the HSE publication, *Noise at Work: A Brief Guide to Controlling the Risks* (INDG362, 2012).

Do I need to employ a consultant to carry out my noise assessment?

No, not necessarily, although the measurement of noise and all the many factors that need to be taken into consideration when carrying out a noise assessment can be complex. If you feel unable to carry out the task yourself or you do not have anyone at your workplace capable of carrying it out then it is probably best to employ a noise consultant.

A consultant will have the necessary noise-measuring equipment to carry out the assessment and will be able to provide you with recommendations on how to tackle any noise issues in the workplace.

Consultants can provide a wealth of knowledge and experience and may be able to make the whole process simple and painless.

Remember

The rules of thumb provided as to whether a noise risk exists should only be used as a guide, and if it is likely that a noise hazard exists in the workplace employers will more than likely need to have a noise assessment carried out by a competent person.

By law, employers are required to assess and identify measures to eliminate or reduce risks from exposure to noise in order that employees' hearing can be protected.

Where the risks are low, the actions taken may be simple and inexpensive, but where the risks are high, they should be managed using a prioritised noise-control action plan.

Remember to review what is being done on a regular basis to ensure that if there have been any changes to the work process or job they are not affecting the noise exposure of employees.

How do I carry out a noise risk assessment?

Firstly, do you have a noise problem?

This depends on how noisy the work environment is and how long workers are exposed to noise.

A noise risk assessment will probably need to be performed if any one of the following workplace criteria applies:

- there are intrusive noises for most of the working day
- employees cannot communicate easily without raising their voices or shouting
- employees cannot hear potential warnings or hazards
- employees use noisy tools or machinery for more than 30 minutes every day
- there are noises due to impacts, explosive sources or guns
- you work in a noisy industry.

Carrying out a noise risk assessment

The aim of a noise risk assessment is to help employers decide what they need to do to protect employees who are exposed to noise.

A noise risk assessment should:

- identify any potential risks resulting from noise
- identify who is potentially affected by these risks
- identify employees who might need health surveillance
- identify what you need to do to comply with legal obligations, (e.g. personal protective equipment, such as earplugs)
- contain a reliable estimate of your employees' exposure, and compare their exposure with the exposure action values and limit values.

It is important that risks are identified to protect employees from being negatively affected by the environment that they are working in.

Estimating employees' exposure

An estimate of the amount of noise that employees are exposed to needs to include:

- the work employees do or are likely to do
- the ways in which employees do the work
- how this work might vary from one day to the next.

The estimate needs to be based on reliable information, such as measurement from the workplace being assessed, measurements from similar workplaces or measurements from machinery suppliers.

All findings from a noise risk assessment need to be recorded, along with an action plan that includes a timetable detailing who is responsible for what work.

The noise risk assessments will need to be reviewed regularly. Any changes in workplace conditions, noise levels, etc. will require an immediate review.

Competence of the risk assessor

It is important that a noise risk assessment is:

- conducted by someone competent in noise risk assessments
- based on advice from someone competent in noise risk assessments.

If no one within the company is competent to carry out noise risk assessments, it will be necessary to go to external consultants to conduct and/or advise on the noise risk assessment.

What are the health effects of noise at work?

Noise at work can cause hearing damage, which may be temporary or permanent. The extent of the damage can lead to a partial loss of hearing or total deafness.

Some people often experience temporary deafness after leaving a noisy place (e.g. after leaving a noisy nightclub or bar). Although this type of hearing loss usually results in the hearing recovering within a few hours, this should not be ignored. Temporary hearing loss is a sign that if the person continues to be exposed to noise their hearing could be permanently damaged.

Permanent hearing damage can be caused immediately by a sudden, extremely loud, explosive noise (e.g. from guns or cartridge-operated machines), but generally it tends to be a gradual process.

It may only be when damage caused by noise over the years combines with hearing loss due to ageing that people realise how deaf they have become.

Hearing loss is not the only problem. People may develop tinnitus (ringing, whistling, buzzing or humming in the ears). Anyone of any age can suffer from hearing damage.

What is a low-noise purchasing policy?

This basically means that when new equipment is bought or hired the employer will ensure that the quietest equipment or machinery is bought or hired. The cost of introducing noise-reduction measures is often reduced quite significantly if quiet equipment can be introduced into the workplace.

What should be included in the low-noise purchasing policy?

The following are a few tips which may be helpful:

- consider at an early stage how new or replacement machinery could reduce noise levels in the workplace – set a target to reduce the noise levels if possible
- ensure a realistic noise output level is specified for all new machinery, and check that tenderers and suppliers are aware of their legal duties
- ask suppliers about the likely noise levels under the particular conditions in which the machinery will be operated, as well as under standard test conditions (noise output data will only ever be a guide, as many factors affect the noise levels experienced by employees)
- only buy or hire from suppliers who demonstrate a low-noise design, with noise control as a standard part of the machine
- keep a record of the decision process that was followed during the buying or hiring of new machinery, to help show that legal duties to reduce workplace noise have been met.

What are manufacturers and suppliers of machinery required to do?

Under the Health and Safety at Work etc. Act 1974 and the Supply of Machinery (Safety) Regulations 2008 (as amended in 2011) a supplier of machinery must:

- provide equipment that is safe and without risk to health, with the necessary information to ensure it will be used correctly
- design and construct machinery so that the noise produced is as low as possible
- provide information about the noise the machine produces under actual working conditions.

New machinery must be provided with:

- a 'declaration of conformity', to show that it meets essential health and safety requirements
- a 'CE' mark
- instructions for safe installation, use and maintenance
- information on the risks from noise at workstations, including:
  - the A-weighted sound pressure level, where this exceeds 70 dB
  - the maximum C-weighted instantaneous sound pressure level, where this exceeds 130 dB
  - the sound power (a measure of the total sound energy) emitted by the machinery, where the A-weighted sound pressure level exceeds 85 dB
- a description of the operating conditions under which the machinery has been tested.

When should hearing protection be used?

If noise cannot be controlled by other methods, such as new machinery, acoustic screening, provision of anti-vibration mounts to machinery or a change in working patterns, then extra protection for employees will be needed. In addition, extra protection may be needed as a short-term measure while other methods of control are being implemented.

Hearing protection should not be used as an alternative to controlling the noise by technical or organisational methods. It should only be used where there is no alternative way of protecting employees from noise exposure.

What are the general requirements for hearing protection?

- If employees ask for hearing protection, it should be provided for them.
- It must be made available for employees to use when the lower action level of 80 dB(A) is exceeded.
- It must be used by employees when the upper action level of 85 dB(A) is exceeded.
- Employers must provide training and information in the correct use of the hearing protection.
- Employers must ensure that any hearing protection that is provided is properly used and maintained.

What should maintenance of hearing protection involve?

Any hearing protection that is provided for employees should be checked and maintained on a regular basis to ensure it remains in good working order. As a minimum the employer should check that:

- it is in good condition and clean
- the seals on ear muffs are not damaged
- the tension on headbands of ear defenders is still good and the band fits well when worn
- employees have not made any modifications to the hearing protection which may mean that it does not provide the correct level of protection
- earplugs are still soft, pliable and clean.

What else can be done?

It is best practice to include the wearing, maintenance and care of hearing protection in the company safety policy.

Managers and supervisors should be encouraged to set a good example by ensuring that they wear hearing protection in areas where it is required.

The hearing protection used must give enough protection – as a guide it should give enough protection for exposure to be below 85 dB.

Employers should ensure that the hearing protection is suitable for the working environment, and consideration should be given to how hygienic and how comfortable it is.

If other personal protective equipment is worn by the employee, such as dust masks or hard hats, the employer should consider how the hearing protection will fit in with the other protection. For example, the wearing of ear defenders may be difficult if the employee also has to wear a hard hat; in this case a hard hat incorporating a set of ear defenders would be required.

It is important to ensure that the hearing protection provided is not:

- designed to cut out too much noise, as this can cause isolation and lead to employees not wanting to wear them
- compulsory to wear, unless the law requires it.

What information, instruction and training do employees need?

Employees need to understand the risk that they may be exposed to from noise in the workplace. If employees are exposed to the lower action level of 80 dB, as a minimum employers should tell them:

- the estimated noise exposure and the risk to hearing
- what is being done to control the risks and exposure – this may include the use of acoustic screening, increased maintenance of noisy pieces of equipment, change in work patterns or provision of hearing protection
- where they can get hearing protection from
- who in the company will be responsible for the provision and maintenance of the hearing protection and who they should report defects to
- the correct way to use the hearing protection, how to look after it, how it should be stored and the areas in which it needs to be used
- details of any health surveillance programme that may be in place.

What is health surveillance for noise?

Health surveillance for hearing usually means:

- regular hearing checks in controlled conditions by a trained professional
- telling employees about the results of their hearing checks
- keeping health records
- ensuring employees are examined by a doctor where hearing damage is identified.

Ideally, health surveillance should be started before employees are exposed to noise, as this will give a baseline that can be used to determine any changes in noise exposure and the effects those changes may be having.

What is the purpose of health surveillance?

Health surveillance is designed to provide the employer with an early warning system as to when employees might be suffering from early signs of hearing damage. It gives the employer an opportunity to do something to prevent the damage getting worse, and ensures that the control measures that are in place are working.

It is important that employees understand the aim and importance of the health surveillance, and that it is there to protect their hearing.

When does health surveillance need to be provided?

If employees are regularly exposed to the upper exposure action level or are at risk for any reason. For example, if an employee is already suffering from hearing loss or is particularly sensitive to hearing damage the employer is required to provide health surveillance in the form of hearing checks.

How often should checks be carried out on employees?

After the initial check, a programme of health surveillance should be implemented, with a regular check being carried out annually for the first 2 years of employment and then at 3-year intervals thereafter. However, this may need to be more frequent if any problem with hearing is detected or where the risk of hearing damage is high.

Who should carry out the hearing checks?

The hearing checks need to be carried out by someone who has the appropriate training. The health surveillance programme needs to be under the control of an occupational health professional (e.g. a doctor or a nurse with appropriate training and experience). The employer is responsible for making sure the health surveillance is carried out properly.

How can health surveillance be arranged?

Larger companies may have access to in-house occupational health services that may be able to carry out the programme. Where there are no facilities in-house an external contractor will need to be used. Further information about occupational health services is available from trade associations and local healthcare services.

What should be done with the results of health surveillance?

The results should be used to make sure that employees' hearing is protected. The records of the health surveillance should be kept and used to provide advice for each employee.

Recommendations should be given by the person carrying out the surveillance and these should be acted on by the employer. The information may need to be used to amend any risk assessments which have been carried out.

Do employers have responsibilities other than to provide ear protection?

Yes. The main responsibility of employers is to reduce the noise *at source* to the lowest level possible.

Noise can be controlled by:

Engineering controls	– purchasing equipment with low noise emissions
	– changing the process (e.g. presses instead of hammers)
	– avoiding metal-to-metal impacts
	– flexible couplings and mountings
	– introducing design dampers
	– correct sizing of ductwork, fans, motors etc.
Orientation and location	– move the noise source away from employees, turning machines around so that noise or sound waves can travel out of the building
	– not putting machines, etc. into hard-surface areas as noise 'bounces' off surfaces
Enclosure	– surround the machine or noise source in sound-absorbing material
	– total enclosure is most effective
	– soundproof the room or work area
	– introduce sound-absorbent materials to surfaces
Use of silencers	– use on ductwork, for motors
	– use on pipes that carry gas, air or steam
	– use on exhaust ventilation systems
Lagging	– lag pipes as an alternative to enclosure
Damping	– dampers can be fitted to ductwork
	– use a double-skin design, preferably with noise-absorbent material in between
Absorption	– acoustic ceiling and wall panels help absorb the sound waves
Screens	– temporary acoustic screens can help reduce levels of noise and these can be moved to where needed
Isolate workers	– remove workers from the noise source by constructing acoustic booths for them to work in.

Usually specialist noise or acoustic consultants will be needed to work out exactly what needs to be done to reduce noise to safe levels.

If you feel that you have 'a din' in the workplace you will need expert advice to reduce noise to tolerable levels.

Who can carry out a noise assessment?

The Noise at Work Regulations 2005 state that it is the employer who must carry out the noise assessment (Regulation 5), but as with all health and safety regulations the employer must ensure that risk assessments are carried out rather than carry them out themselves. The management of Health and Safety at Work Regulations 1999 require employers to appoint competent persons to help them discharge their legal duties and so an employer could appoint an acoustic consultant or general health and safety consultant to carry out the noise assessments. Employers must ensure that the people they appoint are competent and that they understand the working environment in which the noise is created.

A competent person needs to have:

- knowledge
- experience
- information
- interpersonal skills.

In particular, the competent person should show skills appropriate to the situations to be handled, including:

- an understanding of the purpose of assessments
- a good basic understanding of what information needs to be obtained
- an appreciation of his or her own limitations, whether knowledge, experience, facilities or resources
- knowledge of how to make measurements
- knowledge of how to record results, how to analyse them and how to explain them to others
- an understanding of the reasons for using various types of measuring instruments and their benefits
- knowledge of how to maintain, check and attenuate equipment
- knowledge of how to interpret information provided by others and how to assimilate it into other data so as to give overall results of noise levels, etc.

The competent person will *not* need an advanced knowledge of acoustics.

Nor will they necessarily need to know the full details of how to guide and advise the employer on where to obtain further specialist advice.

Competence, as always, is judged in relation to the complexities of the situation to be assessed and should not be overexaggerated.

Assessing noise hazards in industrial production plants that involve many different processes will be more difficult than on a construction site where just a few power tools are in use.

Can someone from within the workforce be appointed as the competent person?

Yes, if they possess the necessary knowledge, skills, experience and information to be able to assess the hazards, risks and control measures necessary in respect of noise at work.

Formal qualifications in acoustics are not necessary except for the most complex of work environments.

A common sense approach to identifying hazards and assessing risks is often more practical than in-depth subject knowledge.

An employee may need to receive further training in risk assessment techniques, the use of sound-level meters, personal dose meters, etc.

Are there any typical topics that an employee or other person needs to be trained in so that they can carry out noise assessments?

Some typical topics for a training course are:

- *Legal requirements*: Information on the Control of Noise at Work Regulations 2005, Management of Health and Safety at Work Regulations 1999 and the Health and Safety at Work etc. Act 1974
- *Purpose of noise assessment*: the need to know how to assess exposure to noise
- *Requirement for noise assessment*: information on sound, noise, sound pressure waves, hazards, risks, etc. and the different types of noise exposure (daily dose levels, dB(A) scales, etc.)
- *Equipment for measuring noise*: sound-level meters, personal dose meters, use of log scales, etc.; the maintenance and attenuation of equipment
- *Noise-measuring procedures*: survey techniques, survey procedures, sampling process, location of microphones etc., measurement of peak noise levels

- *Calculation of noise exposure*: calculation of noise levels over an 8-hour period, peak action levels
- *Noise sources and control measures*: how to control noise, risk reduction, etc.
- *Ear protection*: types and use of ear protection
- *Other sources of information*.

What is a 'hearing protection zone'?

Regulation 7 of the Control of Noise at Work Regulations 2005 states that:

If in any area of the workplace under the control of the employer an employee is likely to be exposed to noise at or above an upper exposure action value for any reason the employer shall ensure that

(a) the area is designated a Hearing Protection Zone
(b) the area is demarcated and identified by means of the sign specified for the purpose of indicating that ear protection must be worn in paragraph 3.3 of Part II of Schedule 1 to the Health and Safety (Safety Signs and Signals) Regulations 1996(1) and
(c) access to the area is restricted where this is practicable and the risk from exposure justifies it

and shall ensure so far as is reasonably practicable that no employee enters that area unless that employee is wearing personal hearing protectors.

A hearing protection zone is one where the noise levels are assessed as high and where they cannot be reduced to safer levels. Employees (and others) are therefore required to wear hearing protection to prevent damage to their hearing.

Employers must ensure that the area designated as a hearing protection zone has suitable signage advising everyone to wear hearing protection.

Signage has to comply with the Health and Safety (Safety Signs and Signals) Regulations 1996.

Construction sites – good practice

Good planning guide for controlling noise on construction sites

A client should include noise-control requirements for both occupational and environmental noise early in the planning stage of a new project. The desired noise-control requirements may be included in a client specification list in the tender document. This can help to avoid unexpected and often very expensive noise control during the construction phase. It allows tenderers to plan how to overcome noise problems in advance.

The client's specifications may include:

- specified noise exposure levels during the construction phase, as per legislative requirements or company policy
- use of quiet or silenced equipment
- adoption of quiet alternative techniques
- use of noise control measures such as silencers, barriers, enclosures
- erection of warning signs identifying noise hazard areas
- time restrictions
- provision of personal hearing protectors and training.

The tenderer's proposal should cover all the client's specifications. The tenderer should prepare a noise-control policy and a noise-control plan to be included in the site-specific safety management plan.

The noise-control plan may be a set of actions required to achieve the noise-control policy and to reduce noise exposure. It may also include information on how the company is planning to meet its obligations, for example:

- a list of equipment to be used – with noise levels at the operator's position and/or at 1 m
- methods undertaken to lower noise exposure (e.g. maintenance, barriers and enclosures)
- restricted hours, rotation of workers in noisy places and special time arrangements (e.g. noisy work done after hours)
- identification of noisy equipment and processes by signs
- the site induction for employees and contractors to include noise levels, noise controls, and correct use and maintenance of personal hearing protectors
- selection and provision of appropriate personal hearing protectors
- audiometric tests.

The main contractor should plan to coordinate subcontractors so that the activities of one do not unnecessarily expose employees of another to noise hazards. It is good practice to nominate one person as the noise coordinator for all noisy activities. Site planning should include:

- preparation of guidance for workers on hazards and the methods to reduce noise
- preparation of schedules of noisy plant and exposure estimates for each phase of work
- laying out the site to separate noisy activities from quieter ones
- scheduling noisy activities to take place when the minimum number of nearby workers are present (out-of-hours noise needs to be carefully planned to avoid neighbourhood annoyance)
- rostering workers to minimise exposure times
- ensuring that workers are well trained, instructed and supervised in noise matters and responsibilities, including the correct use and maintenance of personal hearing protectors.

Once the construction work is in progress, it is essential to monitor the implementation of the noise-control plan. This could be carried out by the client or the main contractor, and could include the following:

- checking if equipment brought onto site complies with specifications (e.g. by obtaining information from suppliers or by undertaking noise assessments)
- reducing noise from identified noise sources by exchanging equipment and/or processes for a quieter alternative or by engineering control methods to quieten the existing equipment or process
- ensuring that all plant is properly maintained (e.g. all noise-control measures such as silencers and enclosures are intact)
- monitoring work schedules to check that noisy work is carried out as specified, away from other workers, outside hours, etc.
- noisy areas that are identified are monitored and well marked so that employees and contractors can avoid entering them unnecessarily
- monitoring to see if training and hearing tests have been carried out, and if personal hearing protectors are adequate and are being worn and maintained correctly
- ensuring that the cause of any hearing loss shown up by audiometry is investigated
- holding toolbox meetings to provide feedback on the effectiveness of noise-control measures and personal hearing protectors to workers, employers and contractors
- posting on safety notice boards the results of any noise assessments conducted and additional noise information.

Typical sound levels (dB(A))

0	Faintest, audible sound
10	Leaf rustling, quiet whisper
20	Very quiet room (e.g. library)
30	Subdued speech
40	Quiet office
50	Normal conversation
60	Busy office
70	Loud radio or TV
80	Busy street in daytime
90	Heavy vehicle close by
100	Band saw cutting metal
110	Woodworking, industrial machine shop
120	Chainsaw
130	Riveting
140	Jet aircraft taking off close by

NOISE RISK ASSESSMENT FORM

Company details: . . . .

Business activity: . . . .

No. of employees: . . . .

Person responsible for noise assessments or competent person: . . . .

Area being assessed: . . . .

Describe work activity: . . . .

No. of employees: . . . .

No., type and use of machines: . . . .

Name of person carrying out assessment: . . . .

Date of assessment: . . . .

Equipment used for assessment: . . . .

Calibration details: . . . .

Model and reference No.: . . . .

Location of noise assessment equipment: . . . .

. . . .

Duration of survey/assessment: . . . .

Exposure assessment measurements

Describe machine types	Noise levels: dB(A)

Maximum exposure times (minutes/hours): ....

Has the first action level been exceeded, i.e. 80 dB(A)?   Yes/No

Has the second action level been exceeded, i.e. 85 dB(A)?   Yes/No

What control measures are in place? ....

....

What control measures are needed in addition to those already in place? ....

....

Does a hearing protection zone need to be designated?   Yes/No

If 'yes', where will it be designated? ....

....

Has signage been displayed?   Yes/No

To what level will noise be reduced once control measures are in place? ....

....

When should remedial measures be completed? ....

....

When should noise assessments be reviewed? ....

....

Other comments (e.g. is health surveillance necessary?): ....

....

....

Date noise assessment completed: ....

Name of assessor: ....

Contact details: ....

Risk Assessments: Questions and Answers
ISBN 978-0-7277-6076-0

ICE Publishing: All rights reserved
http://dx.doi.org/10.1680/raqa.60760.123

Chapter 8
Display screen equipment

What are the requirements of the Health and Safety (Display Screen Equipment) Regulations 1992?

The regulations require employers to minimise the risk arising from working with display screen equipment (DSE) by ensuring that workplaces and jobs are well designed and that equipment is suitable and sufficient and chosen so as not to cause the risk of injury or ill health.

The regulations require every employer to carry out a risk assessment of DSE so that hazards and risks can be identified and control measures implemented.

Workstations have to meet minimum requirements.

Work patterns have to be adapted so that DSE users can have regular breaks away from their screens.

Employers must provide eyesight tests to those employees or users who require them, and must provide corrective spectacles where they are needed for the DSE use.

All employees and other users (e.g. operators) must be given suitable health and safety training and information.

What are the health problems associated with using DSE?

Health problems are not always immediately obvious when using DSE as the symptoms may be quite minor at first, but the repetitive nature of the tasks can exacerbate minor injuries until they become quite debilitating.

Health problems associated with using display screens are:

- upper limb disorders, including pains in the neck, elbow, arms, wrists, hands and fingers
- back ache
- headaches and migraines

- eye strain but *not* eye damage
- fatigue and stress.

Aches and pains to limbs are often referred to as 'repetitive strain injuries' (RSIs), and include carpel tunnel syndrome and tennis elbow.

Do the regulations affect everybody who uses display screen equipment?

No. The regulations in the main apply to the *users* of DSE. The regulations also only apply to employees and the self-employed. Therefore, a 'user' can only be an employee or a self-employed person.

Regulation 1(2d) defines a 'user' as an employee who habitually uses DSE as a significant part of his or her normal work.

In the same regulation, a self-employed person who habitually uses DSE is defined as an 'operator'.

A display screen is not only a computer screen but also television screens, video screens, plasma screens, microfiches screens, etc. Emerging technology is creating new types of screen and it is anticipated that the regulations will cover all of these.

Employers must decide who is a user or an operator under the regulations and apply the requirements of the relevant regulations.

Workers who do not input or extract information from a display screen are generally not users.

Employers need to ask themselves a few searching questions in order to ascertain whether they have users or operators on their staff:

1 Do any of my employees or the people whom I engage as 'contractors' normally use DSE for continuous or near-continuous spells of an hour or more at a time?
2 Do any of them use DSE for an hour or more, more or less daily?
3 Do they have to transfer information quickly to or from the DSE?
4 Do they need to apply high levels of attention and concentration to the work that they do?
5 Are they highly dependent on their DSE to do their job or have they little choice in using it?
6 Do they need special training or skills to use the DSE?

Part-time or flexible workers must be assessed using the same criteria because it is not the length of time they spend 'at work' which counts but the length of time they spend using the DSE.

Sometimes, employers may wish to simplify things and class *all* users of DSE as 'users' or 'operators' under the terms of the regulations. This means that the good practice requirements of the regulations will be applied throughout the organisation.

Examples of display screen users

- Typist, secretary or administration assistant who uses a PC, word processor, etc. for typing documents, etc.
- Word processing worker
- Data entry clerk/operator
- Database operator and/or creator
- Telesales personnel
- Customer service personnel, if computer entry of information is a common part of the job
- Journalists and editorial writers
- TV/video editing technicians
- Micro-electronics testing operators who use DSE to view test results, etc.
- Computer-aided design (CAD) technicians
- Air traffic controllers
- Graphic artists
- Financial dealers

Is it easy to define who is not a user of display screen equipment?
Yes, relatively so.

The answers to the six questions listed previously should enable the employer to easily differentiate who is who.

Anyone who uses DSE occasionally will not be a user under the regulations. Nor will anyone who can choose when or for how long they use DSE.

Laptop users will probably *not* be users as they can (usually) choose when, where and for how long they use their screens and computer.

Receptionists will often not be classed as users as they are not continuously using their screens (unless they predominately operate a switchboard which relies on a screen for extension transfers, etc.).

Are employees who work at home covered by the regulations?

If they are an employee and they use their DSE continuously as part of their job they will be defined as a user irrespective of where they use the equipment.

The display screen and workstation that they use do not have to be supplied by the employer – employees can provide their own equipment but the employer still has to comply with their duties in respect of users and operators and assess the hazards and risks to health.

In order to determine whether home workers are users or operators of DSE the six questions posed earlier will need to be asked for each individual worker.

What are 'workstations' and how do the regulations apply to these?

A workstation is defined in Regulation 1 of the Health and Safety (Display Screen Equipment) Regulations 1992 as:

an assembly comprising:
  (i)   display screen equipment
  (ii) keyboard or other input device
  (iii) optional software
  (iv) optional accessories to the display screen equipment
  (v)   any disk drive, telephone, modem, printer, document holder, work chair, work desk, work surface or other item peripheral to the display screen equipment
  (vi) the immediate work environment around the display screen equipment.

Regulation 2 requires employers to perform a suitable and sufficient analysis of workstations which:

■ are used for the purposes of their undertaking (regardless of who provided them) by users
■ have been provided by them and are used by operators

in order to assess the health and safety risks to which those people are exposed as a result of that use.

The analysis is to assess and reduce risk – it is a risk assessment.

Is it necessary to complete a risk assessment for each workstation and user or operator?

Yes, because each person is different and the effect that using the DSE *may* have on them will be different for each individual. Of course, some individuals may have no ill effects from using a display screen and there will be little you will need to do.

Individual workstations may vary in design, people's tasks will be different and the amount of control they have over their jobs may be different.

The most effective way to conduct risk assessments for DSE users is to create a questionnaire that includes sections on:

- display screens
- keyboards
- mouse or trackball
- software
- furniture
- environment.

The workstation analysis or risk assessment is best done by the individual concerned, once they have had proper training in what they are to look for and how to record the information.

Do the regulations only require an employer to carry out these risk assessments?

No, they are one part of the employer's responsibilities under the regulations.

The regulations themselves require employers to:

- analyse workstations to assess and reduce risks
- ensure that workstations meet minimum specified requirements
- plan work activities so that they include short breaks or changes of activity
- provide eye and eyesight tests on request and special spectacles if needed
- provide information and training.

What are the 'minimum specified requirements' for workstations?

The regulations are quite specific about requirements for workstations and the appropriate regulation, Regulation 3, was amended in 2002 to address a European ruling on the interpretation of the regulation applying to workstations.

The 'minimum specified requirements' apply to all workstations provided by an employer, not just to those used by 'users or operators'.

The European Court in effect stated that all workers, employees or others who use a workstation while at work are entitled to have a workstation that meets the 'minimum specified requirement'.

Do all workstations have to be modified to meet these requirements?

If workstations do not already comply they will need to be modified to meet the conditions laid out in paragraph 1 of Schedule 1 of the DSE regulations.

Paragraph 1 of the schedule lists the following requirements:

■ the components required (e.g. document holder, chair and desk) are present at the workstation
■ that those requirements have effect with a view to securing the health, safety and welfare of persons at work
■ the inherent requirements or characteristics of the task make compliance appropriate.

In effect, employers have to ensure that all workers using DSE have a suitable environment in which to work, have the necessary equipment to work safely, and that the tasks they do are managed effectively so as not to create health and safety issues.

What are the main areas to pay attention to when carrying out a workstation assessment or risk assessment?

Each workstation should be assessed with the following in mind:

■ adequate lighting
■ adequate contrast – no glare or distracting reflections
■ distracting noise minimised
■ legroom and clearances to allow postural changes
■ window covering, if needed to minimise glare
■ software – appropriate to the task, adapted to the user, no undisclosed monitoring of the user
■ screen – stable image, adjustable, readable, glare and reflection free
■ keyboard – usable, adjustable, detachable, legible
■ work surface with space for flexible arrangement of equipment and document, glare free
■ chair – stable and adjustable
■ footrest and/or arm or wrist rest if users need them.

Are all display screen equipment users entitled to an eyesight test and a free pair of glasses?

No. Only those employees who are classed as 'users' under the regulations are covered by the regulation applying to eyesight tests.

An employee who is a user of DSE can request an eyesight test, as can anyone who is to *become* a user, and the employer has to arrange for one to be carried out.

If an existing user requests a test, an employer must arrange for it to be carried out as soon as is practicable after the request, and for a potential user, before they become a user.

The continual use of DSE may cause visual fatigue and headaches, and corrective glasses may reduce the eye strain often experienced. There is no evidence yet available, however, that frequent use of DSE causes permanent eye damage or creates poor eyesight. Users with pre-existing sight conditions may just become a little more aware of them.

Eyesight tests should be carried out by a competent professional and must consider the effects of working with DSE, so the optician (or medical equivalent) will need to know that the eyesight test is for working with DSE.

Once an existing user has had an eyesight test he or she can request a test at regular intervals. The employer should determine what the testing interval is together with the user of the equipment, and should take advice from the optician or other expert.

Eyesight tests that detect short or long sight, eye defects, etc. are *not* the responsibility of the employer – they need only concern themselves with an eye test that addresses any safety or health issues with using DSE.

The employer must arrange for an eyesight test when requested to do so. This could be by having an arrangement with a local optician or by having the test carried out on the premises by a mobile health surveillance unit, etc.

The employer can make arrangements with only one local optician and employees will have no choice who they visit. Alternatively, employers can have employees use their own optician if they prefer. The important thing for employers is that they must facilitate such eyesight tests if requested to do so.

Employers are not responsible for the costs of 'normal' corrective spectacles – these are at the employee's own expense. But an employer is responsible for the cost of any 'special' corrective appliances when the optician has determined that these need to be worn by the user to prevent him or her suffering unnecessary eyestrain while using DSE. The user is only entitled to a basic pair of corrective spectacles necessary for them to continue to use the DSE safely. 'Designer' frames, special lenses, etc. are not the responsibility of the employer.

Employers may make a contribution towards the costs of other types of corrective spectacles if those spectacles include the 'special corrective' features needed for the DSE work.

What does the employer need to do in respect of the provision of training regarding the use of display screen equipment?

The regulations are quite specific about the duties of employers to provide training and information to DSE users.

The employer has to ensure that 'users' and those about to become 'users' of DSE receive adequate health and safety training in the use of any workstation at which they may be required to work.

Training should be provided before a new employee becomes a user of the equipment. The purpose of the training is to ensure that those who are (or will be) users know and understand the hazards and risks associated with using DSE.

Training in the use of DSE can be incorporated in general health and safety training or induction programmes, as it is good practice for everyone to be aware of the hazards and risks associated with all work activities.

It is important that any training programme addresses the steps needed to reduce or minimise the following risks:

■ musculoskeletal problems
■ visual fatigue
■ mental stress.

Managers of those using DSE also need to be trained in health and safety issues relating to DSE as they can have an important influence over a key health hazard – mental stress. Managers must be aware of the legal need for users to take regular breaks away from the screen, for workstations to be ergonomically friendly, etc.

Users must be trained in how to use their DSE effectively. They must know how to make their own personal adjustments to the height and tilt of the screen, screen contrast, etc. They must know when they can take breaks and what other tasks are expected of them.

As with all health and safety training it is important for employers to have a record-keeping system so that they will be able to demonstrate, if called upon to do so, that their employees, users or operators received suitable and sufficient training.

Occupational health issues can take several years to manifest themselves, and elements of musculoskeletal injury may occur after an employee, user or operator has left the company.

Evidence of training can be useful to show that you fulfilled your statutory duties as an employer and that the employee or user was aware of the hazards and risks and knew what to do to control them.

Keep training records for at least 6 years – longer is preferable. Computerised records must comply with the Data Protection Act 1998.

What information do users have to be provided with?

Users of DSE must be provided with adequate information about:

- all aspects of health and safety relating to their workstations
- the steps taken by their employer to ensure compliance with the regulations.

Users and operators of DSE need to know about the risk assessments that have been undertaken, the hazards and risks identified, and the control measures that the employer has put in place to reduce the hazards and risks.

In addition, users and operators need to know what procedures are in place for them to have eyesight tests, the frequency of tests, the provisions for the purchase of 'special needs' spectacles, etc.

Information should be given on when breaks can be taken, what other tasks need to be completed during these times, when they should have training, etc.

Employers should not forget their general duties to *all* employees and others in respect of information, instruction and training on all work activities as required under the Management of Health and Safety at Work Regulations 1999.

Do individual employees have to fill in a self-assessment form?

No, not legally, but it is good practice and allows the employer to have an overview of all employees and any specific and individual concerns they have.

The health effects of using DSE vary from user to user; not everyone reacts the same.

DSE assessments are best done individually. Although generic assessments are not ideal they may be useful to set a standard and provide useful guidance to users.

Individuals' completed self-assessment forms should be collated, reviewed and actioned by the employer or a competent person.

Any individuals who have indicated that they have a specific problem should be reassessed by the competent person and individual control measures agreed.

What are some of the control measures that can be implemented by user themselves?

Practical tips for users include:

- getting comfortable
  - adjust the chair
  - adjust the screen angle
  - adjust the seat height so that eyes are at the same height as the top of the screen and arms are horizontal to the keyboard
  - create enough clear work space
  - remove obstructions and unnecessary equipment
  - adjust the position of the keyboard, mouse, document holder, etc.
  - avoid glare from lights, windows
  - create space to move the feet and legs – provide a footrest if necessary
  - sit comfortably in the chair – make sure it is adjustable
- keying in
  - provide a space in front of the keyboard
  - provide a wrist rest if necessary
  - keep wrists straight
  - do not 'bash' the keys
  - do not overstretch the fingers
- using a mouse
  - sit up straight, do not slouch
  - move the keyboard out of the way
  - keep wrists straight
  - sit close to the desk
  - use a cordless mouse
  - support forearms on the desk
  - do not abuse the mouse – treat it lightly
- reading the screen
  - adjust the brightness and contrast
  - clean the screen regularly
  - use a text size that is as large as practicable on the screen
  - select colours that are easy on the eye
  - do not be afraid to change the screen format to your own preference
  - make sure characters do not flicker and that text, etc. is sharply focused

- posture and breaks
  - move about and change position
  - move the legs and feet
  - take a break and do something different (e.g. answer the phone)
  - do not sit in the same position all the time
  - take frequent short breaks rather than longer infrequent ones.

How does the law on display screen equipment affect the self-employed?

Self-employed people who work in low-risk environments or who undertake low-risk tasks such as office work are no longer required to complete risk assessments or comply with health and safety laws.

If a self-employed person works in a high-risk industry, such as construction, agriculture, with asbestos or genetically modified organisms, then health and safety laws still apply.

Self-employed people carrying out administrative tasks, market research, data inputting, etc. will not need to complete DSE assessments or adhere to the requirements for workstations.

Self-employed people who employ others become an employer, and will be required to comply with all relevant health and safety laws.

I have to hot-desk – do I have to carry out a workstation assessment every time I use a new desk?

It is necessary to ensure that the workstation you are using is not likely to cause you any health risk. In view of this you should, even if you will not be spending a large part of your day there, assess that risk. There are basic things you should always check for, and these are detailed in the HSE booklet *Working with Display Screen Equipment: A Brief Guide* (INDG36(rev4), 2013).

As an employer, it may be useful, if hot-desking is widespread in your organisation, to provide a checklist of what people should assess; this could be attached to the desk or workstation. The self-assessment checklist given at the end of this chapter could be useful for this purpose.

DISPLAY SCREEN EQUIPMENT – SELF-ASSESSMENT

Location/department: ....

Name: ....

Date: ....

1. JOB DESIGN

How long is spent on computer per day? ....

Does this include other roles (e.g. answering telephone)?   Yes/No

Comment: ....

....

Can breaks be taken freely?   Yes/No

Comment: ....

....

2. WORKSPACE AND FURNITURE

Is there sufficient space (3.7 m^2)?   Yes/No

Comment: ....

....

Is desk sufficiently large to allow comfortable arrangement of work?   Yes/No

Comment: ....

....

Is desk height suitable?   Yes/No

Comment: ....

....

Is there adequate light?   Yes/No

Comment: ....

....

Is there excessive noise? Yes/No

Comment: ....

....

Is temperature comfortable? Yes/No

Comment: ....

....

Is there adjustable blinds to windows to prevent reflections? Yes/No

Comment: ....

....

3. EQUIPMENT

Does screen have adjustable controls? Yes/No

Comment: ....

....

Does screen tilt/swivel? Yes/No

Comment: ....

....

Is screen free from reflection/glare? Yes/No

Comment: ....

....

Are digits clear and defined, is screen free from flicker? Yes/No

Comment: ....

....

Is keyboard separate from screen? Yes/No

Comment: ....

....

Is it easy to use/non reflective etc.? Yes/No

Comment: ....

....

Is there adequate space in front of keyboard to support hands and arms? Yes/No

Comment: ....

....

Is a document holder required? Yes/No

Comment: ....

....

4. CHAIR

Is chair fully adjustable? Yes/No

Comment: ....

....

Is it stable? Yes/No

Comment: ....

....

5. OPERATOR

Does operator know how to adjust chair to suit them? Yes/No

Comment: ....

....

Does operator know how to adjust the display and the position of screen
to suit their needs? Yes/No

Comment: ....

....

Are they aware of associated risks? Yes/No

Comment: ....

....

Are they encouraged to take regular breaks? Yes/No

Comment: ....

....

Has information been made available for them regarding visual display
screen use? Yes/No

Comment: ....

....

Signed (auditor): ....

Date: ....

Signed (subject): ....

Case law

Two women employed as data-processing clerks sued British Telecom (BT) for repetitive strain injury.

The system of work they were forced to adopt was long hours spent at the display screen and keyboard, keying in data at high speed. Incentives were given to work faster.

Furniture was not chosen with ergonomics in mind and chairs etc. were not adjustable. No information was given to employees regarding best working posture and practices.

Both women developed painful and ongoing RSI symptoms that prevented them from working.

RSI was a developing industrial disease and information was available about it at the time of the incident (early 1990s).

The two operatives mounted a civil claim for damages and were successful.

The judge held BT liable for the women's injuries as he found that their injuries were purely as a result of their work activity.

BT ought to have taken steps to correct the employees' postures and to have provided proper furniture and workstations, even though the company may not have been fully aware of RSI.

Top tips

- Assess all workstations in the organisation
- Review all equipment for comfort, ease of use, etc.:
  - chairs
  - screens
  - keyboard
  - mouse/trackball
  - lighting/glove
  - environment
  - work/schedule demands
  - software
- Record information in risk assessments
- Decide on control measures to reduce hazards and risks
- Offer eyesight tests to users – do not wait for them to ask
- Determine what level of corrective spectacles you will pay for or towards
- Introduce a comprehensive training programme and give out good levels of information
- Keep training records for as long as possible

Risk Assessments: Questions and Answers
ISBN 978-0-7277-6076-0

ICE Publishing: All rights reserved
http://dx.doi.org/10.1680/raqa.60760.139

Chapter 9
Work equipment

What type of equipment is covered by the Provision and Use of Work Equipment Regulations 1998?

Generally, any equipment that is used by employees while they are at work is covered by the regulations.

Examples of work equipment include:

- photocopiers, printers
- ladders and access towers
- electrical appliances (e.g. kettles)
- knives
- hand tools
- power presses
- drilling and sawing machines
- lifting equipment
- pressure cleaners
- industrial robots
- industrial machinery
- mobile vehicles designed to be used on private land (e.g. dumper trucks and bobcats).

If employees bring their own equipment to use in work (e.g. maintenance tools or hairdressing equipment), it is classed as being work equipment and the employer is responsible for ensuring that it complies with the regulations.

Private vehicles used on public roads are not covered by the regulations; the roadworthiness of vehicles is covered by road traffic acts and other regulations.

Uses of equipment include:

- stopping
- starting

- repairing
- modifying
- maintaining
- servicing
- cleaning
- transporting.

Therefore, maintenance engineers in a factory who only repair the equipment but never actually use it on the production line will nevertheless be classed as a user of the equipment.

What equipment is not covered by the regulations?

The Provision and Use of Work Equipment Regulations 1998 do not apply to equipment used by members of the public, such as:

- compressed-air equipment in a garage
- amusement machines
- vending machines.

However, the general provision of the Health and Safety at Work etc. Act 1974 applies, and employers have to consider the hazards and risks associated with their undertaking in respect of persons who are not their employees.

Do employees have duties under the regulations?

No, employees do not have any specific duties under the Provision and Use of Work Equipment Regulations 1998.

But employees do have duties under Sections 7 and 8 of the Health and Safety at Work etc. Act 1974 not to interfere with equipment given to them by their employer for use at work, or to engage in any reckless acts, etc.

Employees who interfere with work equipment by, for example, disengaging any interlocking guards, will be guilty of an offence under the Health and Safety at Work etc. Act 1974.

Employees have general duties under health and safety law to ensure that they are competent to use work equipment (i.e. that they are adequately trained and have received appropriate information).

What do the Provision and Use of Work Equipment Regulations 1998 require an employer to do?

Employers must comply with the regulations. In particular, they must ensure that work equipment is:

- suitable for use
- suitable for the purpose for and conditions in which it is to be used
- maintained in a safe condition for use so that people's health and safety is not put at risk
- inspected in certain circumstances to ensure that it is and continues to be safe to use
- inspected by competent persons
- subject to written records of inspection and maintenance.

Employers must also ensure that all employees and others (e.g. contractors) using work equipment have received suitable information, instruction and training.

Risks created by the use of work equipment must be eliminated or controlled so that hazards and risks are minimised.

What risks arise from the use of work equipment?

Many things can cause a risk, for example:

- using the wrong equipment for the job (e.g. ladders instead of access towers for work at high levels)
- lack of guards or poor guards on machinery, leading to accidents caused by entanglement, shearing, cutting, trapping, etc.
- having inadequate controls or the wrong type of control so that equipment cannot be turned off quickly and safely or starts accidentally
- failure to keep guards, safety devices, controls, etc. properly maintained so that machines or equipment become unsafe
- failure to provide the right information, instruction and training for those using the equipment.

When identifying the risks think about:

- all the work that has to be done with the machine and other equipment during normal usage and during setting up, maintenance, repair, breakdowns and removal of blockages
- who uses the equipment, including experienced and well-trained employees as well as new starters and those who may have particular difficulties (e.g. impaired mobility)
- employees who may act foolishly or carelessly or make mistakes
- whether guards or safety devices are poorly designed and inconvenient to use or easily defeated, as this may encourage employees to risk injury
- the type of power supply (e.g. electrical, hydraulic or pneumatic) – each type has different risks.

What controls can be implemented when using work equipment?

Hazards can be eliminated or controlled either by taking a number of measures in relation to the machine itself or by following a safe system of work.

Alterations or controls that affect the machine or equipment itself are usually referred to as 'hardware measures'.

Controls that rely on the way people do things are called 'software measures'.

What are some of the 'hardware measures' that can be considered for work equipment?

Guarding

Controlling the risk often means guarding the parts of the machines and equipment that could cause injury.

- Fixed guards should be used wherever possible, and should be properly fastened in place with screws or nuts and bolts which need tools to remove them.
- If employees need regular access to parts of machines and a fixed guard is not possible, use an interlocked guard for those parts. This will ensure that the machine cannot start before the guard is closed and will stop if the guard is opened while the machine is on.
- In some cases (e.g. on guillotines), devices such as photo-electric systems or automatic guards may be used instead of fixed or interlocked guards.
- Check that guards are convenient to use and not easy to defeat, otherwise they may need modifying.
- Think about the best material for guards – plastic may be easy to see through but can be easily scratched or damaged. If wire mesh or similar materials are used, make sure the holes are not large enough to allow access to the danger area. As well as preventing such access, a guard may be used to prevent harmful fluids, dust, etc. from coming out.
- Make sure the guards allow the machine to be cleaned safely.
- Where guards cannot give full protection, use jigs, holders, push sticks, etc. to move the work piece.

Selection and siting of controls

Some risks can be reduced by careful selection and siting of the controls for the machinery and equipment, for example:

- position 'hold to run' and/or two-hand controls at a safe distance from the danger area

- ensure control switches are clearly marked to show what they do
- make sure operating controls are designed and placed to avoid accidental operation (e.g. by protecting any start or stop buttons, pedals, knobs, etc. with covers, flaps or specially designed control boxes, or by designing stop/start functions to operate only when two hands are used (e.g. on tube trains the controls are operated by the 'dead man's handle' – it needs two hands on the controls for the tube to operate))
- interlocked or trapped key systems for guards may be necessary to prevent operators and maintenance workers from entering the danger areas before the machine has stopped
- where appropriate, have emergency stop controls within easy reach, particularly on larger machines, so they can be operated quickly in an emergency.

Before fitting emergency stop controls to machines that have not previously had them fitted, it is essential to check that they themselves will not become a risk. For example, some machines need the power supply to be on to operate the brakes. This power could be lost if the machine is stopped using the emergency stop control.

What are some of the 'software' controls available to employers?

Use and maintenance of hand tools

Many risks can be controlled by ensuring that hand tools are properly used and maintained, for example:

- Hammers: avoid split, broken or loose shafts and worn or chipped heads. Heads should be well secured to the shafts.
- Files: these should have a proper handle and should never be used as levellers.
- Chisels: the cutting edge should be sharpened to the correct angle. Do not allow the head of chisels to spread to a mushroom shape – grind off the sides regularly.
- Screwdrivers: never use these as chisels, and never use hammers on them. Split handles are dangerous.
- Spanners: avoid splayed jaws. Scrap any that show signs of slipping. Have enough spanners of the right size. Do not improvise by using pipes, etc. as extension handles.

Maintenance procedures

Make sure that machinery and equipment are maintained in a safe condition. Controlling the risk often means carrying out preventive checks and maintenance, for example:

- Check what the manufacturer's instructions say about maintenance to ensure that this is carried out where necessary.

- Routine daily and weekly checks may be necessary (e.g. fluid levels, pressures or brake function). When you enter into a contract to hire equipment, particularly a long-term one, you need to establish what routine maintenance is needed and who will do this.
- Some equipment (e.g. cranes) needs preventive maintenance (i.e. servicing) so that it does not break down.
- Equipment such as lifts, cranes, pressure systems and power presses should be thoroughly examined by a competent person at the intervals prescribed by law.
- Make sure guards and other safety devices are routinely checked and kept in working order. They should also be checked after any repairs or modifications.

Carry out maintenance work safely. Many accidents occur during maintenance work. Controlling the risk means following safe working practices, for example:

- Maintenance should be carried out, where possible, with the power to the equipment off and ideally disconnected, or with fuses or keys removed, particularly where access to dangerous parts will be needed.
- Isolate equipment and pipelines containing pressurised fluid, gas, steam or hazardous material. Isolating valves should be locked off and the system depressurised where possible, particularly if access to dangerous parts will be needed.
- Support parts of equipment that could fall.
- Allow moving machines to stop.
- Allow components that operate at high temperatures to cool.
- Switch off the engine of mobile equipment, put the gearbox in neutral, apply the brake and, where necessary, check the wheels.
- To prevent fire or explosions, thoroughly clean vessels that have contained flammable solids, liquids, gases or dust, and check them before hot work is carried out. Even small amounts of flammable material can give off enough vapour to create an explosive air mixture, which could be ignited by a hand lamp or welding torch.

Information, instruction and training
Instruct and train employees. Make sure that employees have the knowledge they need to use and maintain equipment safety, for example:

- give them the information they need (e.g. manufacturer's instructions and operating manuals)
- instruct them on how to avoid risks (e.g. check that the drive is not engaged before starting the engine or machine and do not it use on sloping ground)
- an inexperienced employee may need some instruction on how to use hand tools safely

■ as well as instruction, appropriate training will often be necessary, particularly if the control of the risk depends on how an employee uses the work equipment.

Training may be needed for existing staff as well as inexperienced staff or new starters (do not forget temporary staff), particularly if they have to use powered machinery. The greater the danger, the better the training needs to be. For some high-risk work such as driving fork-lift trucks, using a chainsaw and operating a crane, training is usually carried out by specialist instructors. Remember, younger people can be quite skilful when moving and handling powered equipment but they may lack experience and judgement and may require supervision to begin with.

People who carry out servicing and repairs should have enough knowledge and training to enable them to follow safe working practices. People under the age of 18 years cannot clean any machinery if the act of cleaning exposes them to any risk.

Information, instruction and training needs to be given at regular intervals, the interval depending on the complexity of the equipment being used. People often get complacent and start taking short cuts when using equipment (e.g. avoid using guards), which could expose them to risk.

Training records must be kept for each individual. These should include the date of training and what was covered, and must be signed by the individual.

What needs to be assessed in respect of work equipment?

The Provision and Use of Work Equipment Regulations 1998 require that equipment provided by an employer for use at work is constructed or adapted so as to be suitable for the purpose for which it is provided, and that it is used only for operations and under conditions for which it is suitable.

In addition, the work equipment must be selected with regard to the working conditions, use, hazards and risks, etc. existing on the premises, and must be selected with regard to the health and safety of persons using it.

So, work equipment really needs to be subjected to a risk assessment. In particular:

■ what the equipment is
■ how it works
■ where it will be used
■ who will use it
■ what the hazards and risks are
■ what control measures are needed

- what the residual risks are
- what training operatives will need.

Employers must protect employees from any dangerous parts of machinery.

What are some of the hazards and risks associated with using work equipment?

Work equipment is a major cause of injury and has contributed to fatalities, major injuries, 'over 7-day injuries', lost time and near-miss incidents.

Failing to use equipment safely increases the risks of injury.

Hazards include:

- entanglement
- entrapment
- electricity
- hot surfaces
- impact damage
- excessive noise
- fire and explosion.

Risks include:

- death
- serious injury
- amputation of limbs
- electric shock
- musculoskeletal problems
- industrial disease (e.g. deafness).

When should a risk assessment be carried out for work equipment?

A risk assessment is essential when new equipment is to be brought into the workplace, when equipment is to be changed or adapted, and when the work process changes.

A risk assessment should also be carried out if the use of the equipment will affect people other than those at work.

Under the Dangerous Substances and Explosive Atmospheres Regulations 2002, a risk assessment must be carried out if any work activity or use of equipment involves or will involve an explosive atmosphere (e.g. the use of electrical equipment in a flammable environment).

What are the key things I need to know about the Lifting Operations and Lifting Equipment Regulations 1998?

Lifting equipment, including lifts, hoists, eye bolts, chains, slings, etc., and lifting operations are covered by these regulations.

Lifting equipment must be adequate in strength and stability for each load.

All equipment used for lifting persons must be safe so that people cannot be crushed, trapped, struck by or fall from the lifting carrier.

Where safety ropes and chains are used they must have a safety coefficient of at least twice that required for general lifting operations.

Lifting equipment must be installed or positioned in such a way that it reduces the risk of the equipment striking a person or of the load drifting, falling freely or being released unintentionally.

Suitable devices must be available to prevent persons from falling down a shaft or hoist way.

Equipment used for lifting, including that used for lifting people, must be marked with safe working loads.

If equipment is not suitable for lifting persons it must be marked accordingly.

Lifting operations must be properly planned, appropriately supervised and carried out in a safe manner.

Lifting equipment must be regularly inspected and tested and a report of its condition produced:

- before being used for the first time
- after assembly and being put into service for the first time at a new site.

Lifting equipment used for lifting *people* must be examined and tested every *6 months*.

All other lifting equipment (e.g. goods hoists) must be examined and tested every *12 months*, unless the competent person deems it to need inspecting more frequently.

Records must be kept, although these can be electronic.

Persons carrying out examination and testing must be competent.

What are some of the hazards, risks and control measures needed in respect of lifting equipment?

The Lifting Operations and Lifting Equipment Regulations 1998 require that lifting equipment is inspected by a competent person either:

- annually, if only goods are lifted
- twice a year, if people are lifted.

The regulations also require lifting operations to be assessed to ensure that the equipment is safe and suitable for the tasks in hand.

Lifting equipment includes lifts, lift cars and cages, lifting devices, hoists, block and tackles, ropes, chains, eye bolts, wheelchair lifts, scissor lifts, cradles and cherry pickers.

Some common hazards to watch out for are:

- mobile cranes might overturn if not properly set up with stabilisers
- lifting machinery may fail through mechanical fault or damage or through overloading
- dangerous parts of machinery (e.g. closing traps), especially in scissor lifts
- trapping risks between the lifting machinery and overhead structures
- overhead electrical lines or cables close to lifting machinery.

Typical control measures are:

- lifting equipment is maintained in a safe condition
- lifting equipment is examined, tested and certificated in accordance with statutory requirements
- up-to-date records are kept
- the safe working load (SWL) is clearly marked on all equipment
- lifting equipment is always visually inspected prior to use
- correct equipment is used for the task
- the SWL is not exceeded
- lifting equipment is only used, inspected and maintained by a competent person(s)
- hydraulic hoists are supplied by a pre-vetted contractor
- contractors supply a risk assessment that applies to the task in hand
- consider medical review of employees for fitness to work at heights.

Are hand tools considered equipment under the Provision and Use of Work Equipment Regulations 1998?

Yes. Hand tools can cause a number of injuries if not used correctly, and although they may be small they can still be dangerous.

A risk assessment should consider the hazards, who could be harmed and how, what needs to be done to control the risks, etc.

Hand tools could include knives, hammers, chisels, screwdrivers, woodworking tools, maintenance kits and electrical appliance testers.

Common hazards are:

- handles in poor condition can cause hand injuries
- mushroomed heads on steel chisels can cause fragments to be ejected
- blunt cutting edges can require excessive force to be used
- improper use of knives (e.g. cutting towards the body)
- hand tools that are poorly modified or used for purposes they are not designed for.

Typical control measures:

- good quality tools are provided and maintained
- users are adequately trained and competent
- special purpose tools are provided where necessary (e.g. parcel knives for opening packages)
- sharp knives are fitted with shielding or retracting blades where possible.

What are the hazards and risks to be assessed with regard to workshop equipment?

Workshop equipment will include lathes, milling machines, circular saws, power presses, grinding machines, drilling machines and a whole range of other industrial-type equipment (e.g. abrasive wheels).

Woodworking machinery and power presses are known to be particularly high-risk pieces of work equipment, and need special attention in respect of health and safety when they are used by people at work.

Power presses must be inspected by competent persons and records must be kept on a *daily* basis to ensure that they are safe to use. In addition, power presses need to be thoroughly examined by a competent person at specified intervals.

Common hazards associated with workshop equipment include:

- inadequate guarding allows contact with dangerous parts (e.g. rotating parts, cutting blades and traps between closing tools)
- improper access into machinery cabinets by unauthorised staff (e.g. during fault finding)
- ejection of materials from machinery
- swarf and dust are allowed to build up
- controls that are poorly identified and positioned
- harmful substances and dusts emitted during machining.

Typical control measures are:

- suitable guarding is provided and maintained
- machine is used only by trained competent people
- shields, guards and protective equipment are provided where necessary
- the machine is left fit for next use
- machine controls are clearly identified and conveniently positioned
- good general ventilation and local exhaust ventilation are provided and used.

What are the hazards and risks to be assessed with regard to pressure systems?

The Pressure Systems Safety Regulations 2000 apply to pressure vessels and systems, and are designed to ensure that potentially dangerous and explosive equipment is used and maintained in a safe condition.

The hazard from a pressure vessel is high – it could explode and cause mass injury, fire and great consequential damage. However, provided a pressure system is well maintained the risk could be quite low.

Pressure systems include:

- boilers
- steam heating systems
- air compressors
- pressure cookers
- steam ovens
- heat exchangers
- process plant and pipes
- pressure gauges and hoses
- refrigeration plant.

Common hazards include:

- poor equipment and/or system design
- poor maintenance of equipment
- unsafe systems of work
- operator error, and poor training or supervision
- incorrect or poor quality installation
- inadequate repairs or improper modifications.

Typical control measures are:

- safe and suitable equipment is provided
- the operating conditions are known (e.g. what fluid or gas is being contained, stored or processed)
- the process conditions are known (e.g. pressure and temperature)
- the safe operating limits of the system (and any equipment directly linked to or affected by it) are known
- there is a set of operating instructions for all of the equipment and for the control of the whole system (including emergencies)
- appropriate employees have access to these instructions and are properly trained in the operation and use of the equipment or system
- suitable protective devices are fitted and are kept in good working order at all times
- checks are made to ensure that the protective devices function properly and are adjusted to the correct settings
- warning devices are fitted that are noticeable either by sight or sound
- the system and equipment are properly maintained, taking account of its age, the environment and its use
- when protective devices have to be isolated for maintenance, alternative arrangements are made to ensure safety levels are not exceeded without detection
- there is a safe system of work to ensure that maintenance work is carried out properly and under suitable supervision
- appropriate training is given to everybody operating, installing, maintaining, repairing, inspecting and testing pressure equipment
- all persons carrying out the work are competent
- the equipment has been examined and a written scheme of examination has been prepared by a competent person.

What must an employer do when providing mobile work equipment?

Mobile work equipment, including self-propelled, remote-controlled work equipment, is subject to specific requirements in addition to the requirements for normal work equipment. These requirements cover:

- the suitability of equipment used for carrying people
- the minimising of rollover risk
- the provision of equipment to restrain or protect people (in the event of rollover or overturn)
- the control of the equipment, including operator vision and lighting (where required for safety).

When providing mobile work equipment, you must ensure that:

- where employees are carried (e.g. elevating platforms), the equipment is suitable for that purpose
- the risks from rolling over are minimised, and any person being carried is protected in the event of fall or rollover; this should include protection against crushing, through the provision of a suitable restraint and a rollover protection system
- self-propelled equipment can be controlled safely with braking devices, adequate driver vision and, where necessary, lighting
- adequate guarding is in place to prevent any risks from drive shafts that power accessories attached to mobile work equipment.

What other legislation applies to the use of equipment at work?

The Lifting Operations and Lifting Equipment Regulations 1998 apply to equipment that is used to lift materials or people, covering such equipment as:

- cranes
- fork-lift trucks
- lifts
- hoists
- mobile elevating work platforms
- scissor lifts
- disabled person lifts
- tail gates (hydraulic, electric, etc.)
- vehicle inspection platforms.

It is necessary to apply both the Lifting Operations and Lifting Equipment Regulations 1998 and the Provision and Use of Work Equipment Regulations 1998 to the same piece of equipment, as they cover different aspects of equipment safety.

Do the Provision and Use of Work Equipment Regulations 1998 apply to equipment that is hired?

Yes. As an employer, if you hire equipment then you must comply with the regulations when the equipment is in use at the workplace.

If you hire out equipment you are responsible for ensuring that the equipment is safe to use at the point of hire.

Persons who hire out equipment should make reasonable enquiries of the hiree to ascertain what the equipment will be used for and give reasonable advice on how it should be used.

The safe *use* of the equipment hired is the responsibility of the person who hires it.

What controls should be put in place when using fork-lift trucks?

The most important control to place on the use of fork-lift trucks is a 'permit to use' system that restricts the use of trucks to trained and competent persons.

Additional controls to prevent accidents are:

- always use the correct fork-lift truck for the task
- ensure the braking system is adequate
- ensure operators, supervisors and managers are adequately trained
- lay out the site to ensure the fork-lift truck can move safely without danger to pedestrians
- remove any obstructions, where possible, or ensure they are clearly marked
- fit seat restraints, where appropriate
- fit visibility aids, such as mirrors, where appropriate.

Additional controls when operating a fork-lift truck are:

- do not overreach or overbalance
- avoid travelling on uneven or steeply sloping ground
- do not travel too fast, in particular around corners
- do not overload
- lower the load before operating the truck
- ensure adequate visibility to avoid collisions with pedestrians and objects
- protect obstacles such as support columns, pipework or other plant with impact barriers
- ensure each operator has site-specific instructions
- ensure that the fork-lift truck is inspected and serviced at appropriate intervals and thoroughly examined by a competent person at least every 12 months or at intervals set by the competent person.

Do woodworking machines fall under the jurisdiction of the Provision and Use of Work Equipment Regulations 1998?

Yes. Woodworking machines are included in the definition of 'equipment'.

The Woodworking Machines Regulations 1974 were repealed by the Provision and Use of Work Equipment Regulations 1998.

A specific approved code of practice, *Safe Use of Woodworking Machinery* (L114, 2nd edn, 2014), has been published by the Health and Safety Executive (HSE), and all employers who use woodworking machines should refer to the guidance given.

Woodworking machines should be operated by trained, competent people, be provided with guards and adequate stop and start controls, and be safely sited on level ground, in good lighting with adequate ventilation in the work area, etc. Operatives should have suitable space in which to work, and must be provided with push sticks or other devices to keep hands away from blades.

An important control to introduce when using woodworking machines is dust collection. Wood dust is harmful to health, and operators must be protected from inhaling the fine particles.

Machines must be provided with efficient means of collecting dust or chippings (e.g. local exhaust ventilation).

What records of maintenance need to be kept?

Employers must show that they have carried out their duties under the Provision and Use of Work Equipment Regulations 1998, and any maintenance of work equipment must be recorded.

Routine checks and regular maintenance checks must be recorded so that dates, times, who took action, what needed to be done, when remedial works were completed, etc., are easy to see.

Remember, employers have a duty to maintain work equipment in a safe condition so that people's health and safety is not put at risk. Records of how you have discharged that duty will be essential in the case of an accident.

Does each piece of work equipment need a risk assessment?

A risk assessment will be needed for each type of equipment that is used at work. If a group of similar pieces of work equipment is used (e.g. hand-held carpentry tools) then one risk assessment could cover all the tools.

The task undertaken should be described together with the equipment being used, and the hazards and risks associated with the use of the equipment must be identified. While users are usually well covered by the risk assessment, consideration must be given to any persons in the vicinity who could be affected by the work equipment (e.g. people in warehouses where fork-lift trucks are used).

The more dangerous the piece of equipment the more detailed the risk assessment.

How do the Provision and Use of Work Equipment Regulations 1998 apply to power presses and mobile work equipment?

Power presses and mobile work equipment are considered potentially very dangerous pieces of equipment if not properly maintained and used. The Provision and Use of Work Equipment Regulations 1998 contain additional requirements for power presses and mobile work equipment, and employers would be wise to consult more detailed information available from the HSE or local authority.

Mobile work equipment includes dumper trucks, fork-lift trucks, mobile elevating work platforms, scissor lifts, cherry pickers, etc.

Any mobile work equipment designed to carry people must meet specific requirements, such as having roll-over protection and good stability.

Power presses have the potential to cause serious accidents, and must therefore have appropriate guards or protection devices to prevent entrapment and impact injuries. All guards or protection devices must be inspected at specified intervals and the condition recorded. In addition, there must be daily inspections to ensure that the guarding is working properly.

Can any employee inspect work equipment and record the details in the inspection log?

No. Any person who inspects work equipment must be competent to do so.

Parts of the Provision and Use of Work Equipment Regulations 1998 require an 'appointed person' to be nominated to inspect equipment (e.g. the guards and safety devices used on power presses must be inspected daily by an appointed person).

A 'competent' person is anyone who has the knowledge, experience and training to undertake the tasks allocated, so a competent person who inspects equipment needs to know what the equipment is used for, how it is used, where and when it is used, what maintenance is needed, where to identify faults, who uses the equipment, etc.

An 'appointed' person is someone who has been adequately trained and who is competent to work on each power press.

Whoever inspects work equipment must be able to identify faults and assess how those faults will affect the safe use of the equipment, and must be able to identify who might be harmed and how if they were to continue using faulty equipment.

What is 'thorough examination and testing'?

Certain pieces of equipment, such as power presses and mobile work equipment, need to be subjected to a thorough examination and test by a competent person.

'Thorough examination and testing' is exactly as it sounds – a comprehensive assessment of the equipment to confirm that it is working as it should be, that all guards and safety devices are working properly, etc.

Insurance companies will often require written evidence of the thorough examination and testing in order to continue providing insurance cover for the equipment.

The Provision and Use of Work Equipment Regulations 1998 and other relevant regulations such as the Lifting Operations and Lifting Equipment Regulations 1998 permit the competent person to determine the frequency of thorough inspection and testing unless set intervals are prescribed in legislation (e.g. passenger lifts must be inspected at least every 6 months, although they can be inspected more frequently if the competent person determines it necessary).

Is there anything I should know or do when ordering or buying new equipment?

All new equipment must be suitable for the job you intend it to do. It would be inappropriate, for instance, to buy a domestic step ladder for use in industrial or commercial buildings because a domestic step ladder is not designed to take the extra weight and constant use that such a piece of equipment would be subjected to in an industrial or commercial organisation.

All new equipment has to comply with safety law regarding design and manufacture.

You should look for kite marks, CE numbers, certificates of conformity and general statements from the suppliers that they meet international standards.

If the equipment generates noise, ensure that there is a noise assessment statement and information on decibel levels. A loud item of equipment may put you in breach of the Control of Noise at Work Regulations 2005.

Be wary of cheap imports that do not have associated documentation. If the machine is faulty and your employee has an accident, you will still be liable.

What should employees do to ensure the safe use of equipment?

Employees should check that:

- they know how to use the machine
- they know how to stop the machine before they start it
- all guards are in position and all protective devices are working
- the area around the machine is clean, tidy and free from obstruction
- they are wearing appropriate clothing and equipment, such as safety goggles or shoes, where necessary
- they are not wearing inappropriate clothing, jewellery or have long hair which could be caught up in machinery.

They should:

- tell the supervisor at once if they think a machine is dangerous because it is not working properly or any guards or protective devices are faulty
- stop using the machine until the matter has been checked.

They should never:

- use a machine unless they are trained to do so
- try to clean a moving machine if this could be dangerous – they should switch it off and unplug it or lock it off
- use a machine or appliance that has a danger sign or tag attached to it – danger signs should be removed only by an authorised person who is satisfied that the machine or process is safe
- wear dangling chains, loose clothing, gloves or rings, or have long hair which could get caught up in moving parts
- distract other people who are using machines, fool around or deliberately misuse the equipment
- allow passengers to be carried on vehicles such as dumper trucks or fork-lift trucks unless the vehicle is designed for such use.

Case studies

Building cleaning and maintenance

An operative of a high-pressure water jetting company suffered a serious leg injury when a water-jetting lance he was using failed to shut off safely. The operator was engaged in jet washing the external façade of the building.

At the end of the period of jetting (high-pressure washing) the operator released the trigger on the jetting lance, which should have tripped the pump supplying the jetting water at very high pressure. The pump trip should have opened a sump valve that relieved the pressure of the water in the hose. However, the sump valve failed to open, leaving the hose pressurised.

The operator knew that something had gone wrong but decided to investigate and pulled the lance trigger, causing high pressure water to lacerate his legs.

The investigation found that the manufacturer of the pressure-washing equipment had supplied a faulty machine and that the safety lock-off valve was missing. The manufacturer was prosecuted. In addition, the company employing the operator was prosecuted for failing to implement a safe system of work, failing to have a risk assessment and failing to give appropriate information and training to the operator on how to use work equipment.

Use of equipment in a kitchen

A major UK restaurant operator was fined for failing to ensure that employees were competent and trained in using a gravity-feed slicer in the kitchen. In addition, the operator had failed to undertake risk assessments and to keep maintenance and training records, and had turned a blind eye to the fact that a young person was using and cleaning a dangerous machine.

The local environmental health officer investigated as a result of an accident report under the Reporting of Injuries, Diseases and Dangerous Occurrences Regulations 2013. An employee had suffered severe lacerations to a hand.

The employer should have put proper procedures in place to ensure that all employees were properly trained and aware of the risk assessment. The machine was never checked to ensure that all the blade guards were in place.

If the employee had been properly trained the accident would not have happened.

WORK EQUIPMENT – CHECKLIST

Name of business: . . . .

Location of equipment: . . . .

Type of equipment under assessment: . . . .

. . . .

Who uses it: . . . .

Subject	Yes	No	N/A	Actions required
User competency				
Have all operatives received suitable information, instructions and training?				
Are training records kept?				
Use of equipment				
Is equipment being used in accordance with the manufacturer's instructions, the company's safety policy or other safety procedures?				
Are all guards that should be used in place?				
Are operatives following the safe system of work and using appropriate controls?				
Does the use of the equipment pose any hazards to the operative?				
Is noise controlled?				
Are any exhaust fumes, dust, etc. suitably controlled?				
Are suitable warning notices supplied on the equipment and are they clearly visible to all users?				

Subject	Yes	No	N/A	Actions required
Maintenance				
Are there suitable maintenance records for the equipment?				
Does any part of the equipment appear to be defective (e.g. broken guards, frayed leads, broken casing)?				
Are operatives aware that they need to check the maintenance records?				
Specific hazards				
Is protection provided in relation to: ■ items falling ■ items being ejected ■ overturning ■ collapse ■ overheating ■ fire ■ disintegration ■ explosion ■ unscheduled start				
Environment				
Is there adequate lighting?				
Is good housekeeping practiced?				
Is local exhaust ventilation provided?				
Is the work area free from hazards?				
Is noise controlled?				
Is the work area overcrowded and are operatives at risk?				
Detail any other issues				

Subject	Yes	No	N/A	Actions required
Gas/electricity				
Does the equipment use gas or electricity?				
If yes, are procedures in place to ensure safety from gas release, carbon monoxide, etc., or electric shock?				
Fire safety				
Does the operation of the equipment pose any specific fire safety risk to the operatives or overall work area?				
Dangerous machinery				
Is the machine classified as a dangerous machine and is it subject to specific legal requirements (e.g. Lifting Operations and Lifting Equipment Regulations 1998)?				
Are young persons prevented from using or cleaning dangerous machines?				

ACTION SUMMARY

Please describe in more detail the steps which need to be taken to ensure that the equipment will be used safely. Indicate who should action any tasks and in what timescale.

Actions	Who	By when

Signed: . . . .

Person carrying out the check/inspection: . . . .

Date of check/inspection: . . . .

Equipment needs to be reviewed on: . . . .

WORK EQUIPMENT – RISK ASSESSMENT FORM

Task/activity/equipment

What are the hazards or dangers?	Risk	Groups at risk	Probable severity	Current controls	Further precautions or site-specific control measures required	Likelihood of harm	Total risk rating

WORK EQUIPMENT – RISK ASSESSMENT FORM Continued

Probable severity		Likelihood of harm		Risk rating
Trivial injury = 1		Very unlikely = 1		1 to 6 = low
Minor injury = 2		Unlikely = 2		
Moderate injury = 3	×	Possible = 3		7 to 14 = medium
Major injury = 4		Likely = 4		
Dangerous contamination = 5		Very likely = 5		15 to 36 = high
Fatality = 6		Extremely likely = 6		

Assessor	
Date	
Review date	

Details of reviews or amendments

Date of review/amendment	Carried out by	Comments

WORK EQUIPMENT – EXAMPLE RISK ASSESSMENT

Task/activity/equipment	Use of pressure washers

What are the hazards/dangers?	Risk	Groups at risk	Probable severity	Current controls	Further precautions or site-specific control measures required	Likelihood of harm	Total risk rating
Faulty electrical equipment	Electric shock	Employees conducting the activity	4	Equipment to be used in accordance with manufacturer's instructions Electrical checks to be carried out – visual checks of leads and plugs Annual PAT testing to be carried out by a competent person Any damaged equipment taken out of use and repairs only undertaken by a competent person		3	Medium

WORK EQUIPMENT – EXAMPLE RISK ASSESSMENT Continued

What are the hazards/dangers?	Risk	Groups at risk	Probable severity	Current controls	Further precautions or site-specific control measures required	Likelihood of harm	Total risk rating
Pressure of water	Water-pressure 'burns'	Employees conducting the activity	3	Manufacturer's representatives to give training to employees Manufacturer's instruction to be readily available at each premises Pressure washer has a safety valve/switch to prevent pressure build up and potential explosion Pressure washer to be used for cleaning floor and wall surfaces only		3	Medium
Trailing leads	Slips, trips, falls Personal injury	Employees conducting the activity Anyone in the vicinity	3	No trailing wires from pressure washers to be left across walkways or corridors when the premises are occupied, in order to reduce the risk of tripping		3	Medium

What are the hazards/dangers?	Risk	Groups at risk	Probable severity	Current controls	Further precautions or site-specific control measures required	Likelihood of harm	Total risk rating
Flooding of area causing slips, etc.	Personal injury	Employees conducting the activity Anyone in the vicinity	3	Employees trained to clean up surplus water Employees trained to identify spillages and use hazard signage Spillages mopped up immediately		3	Medium
Water and electricity	Electric short-circuits leading to fires	Employees conducting the activity	6	Equipment not to be used near live electrical equipment – appliances to be switched off and unplugged or electrical connections protected		2	Medium

WORK EQUIPMENT – EXAMPLE RISK ASSESSMENT Continued

Assessor	
Date	
Review date	

Probable severity		Likelihood of harm		Risk rating	
Trivial injury = 1		Very unlikely = 1		1 to 6 = low	
Minor injury = 2		Unlikely = 2			
Moderate injury = 3	×	Possible = 3		7 to 14 = medium	
Major injury = 4		Likely = 4			
Dangerous contamination = 5		Very likely = 5		15 to 36 = high	
Fatality = 6		Extremely likely = 6			

Details of reviews or amendments

Date of review/amendment	Carried out by	Comments

WORK EQUIPMENT – MAINTENANCE LOG

Equipment description	Maintenance carried out	When	By whom	Actions

Risk Assessments: Questions and Answers
ISBN 978-0-7277-6076-0

ICE Publishing: All rights reserved
http://dx.doi.org/10.1680/raqa.60760.171

Chapter 10
Working practices, lone workers and safe systems of work

What is 'lone working'?

Lone workers are those who work by themselves without close or direct supervision.

They are:

- people in permanent premises where only one person works on the site at any one time (e.g. in small workshops, kiosks, shops and home workers)
- people who work separately from others in factories, warehouses, research and training establishments, leisure centres or fairgrounds
- people who work outside normal hours as cleaners, security, special production, night-shift workers, or maintenance and repair staff
- people who work away from their home base on construction sites, in plant installation, maintenance, cleaning work, electrical repairs, lift work, painting and decorating, or vehicle recovery
- agricultural and forestry workers
- service workers who collect rents, postal workers, home helps, community nursing staff, pest-control workers, drivers, engineers, estate agents, sales representatives and similar professionals visiting domestic and commercial premises.

What are the legal duties and responsibilities employers have for lone workers?

Health and safety law does not prohibit lone working and there are no specific regulations that apply to lone working. General health and safety law applies.

However, there can be hazards and risks to employees from lone working and employers must carry out a risk assessment as part of their broad duties under the Health and Safety at Work etc. Act 1974 and the Management of Health and Safety at Work Regulations 1999.

These regulations require the identification of the hazards found at work, assessing the risks arising from these hazards, and then putting measures in place to control the risks.

How should an employer assess and control the risks of lone working?

A risk assessment should indicate any significant risk, and detail how the risks should be adequately controlled for lone working to continue.

Risk assessment often identifies the correct level of supervision or backup required. Some risk assessments, such as those for working in confined spaces, state that communication and rescue arrangements need to be in place where at least one other person needs to be present.

Control measures may include training, instruction, communications and supervision.

Personal protective equipment (PPE) may be necessary.

If a risk assessment shows it is unsafe to work alone, then arrangements should be in place for providing help or backup.

If the employee is at another employer's workplace, the occupier should inform the lone worker's employer of the risks and of control measures needed.

For organisations with five or more employees, the risk assessment of significant findings must be recorded.

What are examples of safe arrangements for lone workers?

Safe working arrangements for lone workers are no different from organising the safety of other employees:

- It must be identified if the lone worker can adequately control the risks of the job.
- Precautions must be in place for both normal work and for emergencies such as fire, equipment failure or sudden illness.

Other considerations:

- Does the lone worker have a safe way in and out of the workplace?
- Can one person handle temporary access equipment, plant, goods or substances?
- Is there a risk of violence?
- Are women especially at risk?
- Do young people work alone?

Lone workers in many situations also face greater risks from violence and aggression.

Are there any rules about the medical suitability of lone workers to carry out their jobs?

Employers must check that lone workers have no medical condition that would make them unsuitable for working alone, seeking medical advice if necessary.

What training must be provided for lone workers?

Training is particularly important where there is limited supervision:

- Lone workers need to be sufficiently experienced to fully understand the risks and precautions required.
- Employers should set limits of what may and may not be done while working alone.
- Lone workers should be competent to deal with unusual or new circumstances beyond their training, and know when to stop and seek advice.
- Employees must know how to respond to emergency situations.

What supervision of lone workers are employers required to provide?

The extent of supervision depends on the risk and the ability of the lone worker to identify and handle health and safety issues.

Employees new to a job may need to be accompanied until competencies are achieved. Employers must be confident that they have set down the core competencies that have to be achieved. Supervisors may periodically visit to observe the work being done.

What are some examples of usual control measures that employers put in place?

Employers must determine the control measures that are suitable for the lone work to be carried out, and, in general, generic risk assessments will not be deemed suitable and sufficient risk assessments.

Should generic risk assessments be considered, some common control measures include:

- regular contact by radio, telephone or mobile phone
- automatic warnings, which should be activated if specific signals are not received at base
- other warnings that raise an alarm in the event of an emergency
- checking that the lone worker has returned to base, or home, on completion of the work.

What needs to be considered in respect of emergencies and lone working?

All employees and others should be able to respond effectively to any emergency situation:

- Lone workers should be capable of responding correctly to an emergency.
- Emergency procedures should be in place with the worker trained to respond.
- Lone workers should have access to a first-aid kit or facilities.
- Risk assessment may indicate that the lone worker needs first aid training.
- Consideration may need to be given to health surveillance for lone workers (e.g. checking blood pressure levels and general health so that any health conditions can be identified and the employee given the opportunity to seek alternative job tasks).

What responsibilities do lone workers have for themselves?

Lone workers, including self-employed ones, have a responsibility to:

- take reasonable care to look after their own health and safety
- safeguard the health and safety of other people affected by their work
- cooperate with their employer's health and safety procedures
- use tools and other equipment properly, according to relevant safety instructions and any training they have been given
- not misuse equipment provided for their health and safety.

Should an employee fail to follow the procedures laid down by their employer they could be at risk of prosecution under Section 7 of the Health and Safety at Work etc. Act 1974 if an accident or incident were to occur.

What about the safety of lone workers working from home?

Do not assume that employees who work at home are not at risk. As an employer, you have the same responsibility for the health and safety of people who work from home as for any of your other workers.

Ensuring premises and work practices are safe

In many cases, domestic premises will not be as well equipped as business premises that have been built specifically as work environments. For example, a lone worker's house may have poor lighting, ventilation and equipment, or its electrical wiring may be old and unreliable.

It can also be difficult to ensure that homeworkers work in a safe way. For example, it is difficult to check that regular breaks from working at a computer are being taken or that possible distractions such as telephones, radios and televisions are not increasing the risk of an accident occurring.

Employees should be given appropriate training in the potential hazards of working at home, with the examples given above included in the training plan.

Protecting the health and safety of lone homeworkers

The health and safety risk assessment must consider whether work being done at home might cause harm, either to the homeworkers themselves or to other people. This may require visiting the homes in question, although a thorough questionnaire or similar process may be able to identify key potential hazards.

Consider drawing up a homeworking policy that sets out key steps to be taken by people working at home to protect their health and safety.

You may also want to insist that certain safety standards are met before allowing people to work from home.

How should I monitor lone workers' health and safety?

It is not possible to continuously supervise lone workers, but communicating with them regularly and monitoring their working conditions and practices plays an important part in reducing health and safety risks.

Employers are required to consult their workforce on health and safety matters, either directly or via worker representatives. It is important to talk to employees, as they are a valuable source of information and advice. Effective consultation will also help to ensure that all relevant hazards are identified and appropriate, proportionate control measures chosen.

There are various steps you can take depending on the type of work being carried out and the type of premises being used. You could:

- Make regular visits to a lone worker. This is the best way of monitoring workplace hazards and safe working practices.
- Consider increased supervision where employees are new to a job, undergoing training, doing a job that presents special risks or dealing with new situations.
- Set up a simple procedure for lone workers to report incidents such as accidents and near misses.
- Make sure lone workers know that they should take regular breaks and avoid working excessively long hours.

- Ask people working on their own if they feel there are any safety concerns that are not being addressed.
- Encourage lone workers to seek help and advice if any safety concerns arise.
- Encourage lone homeworkers to visit your business premises every now and again. Face-to-face contact with colleagues might help them feel part of a team.
- Make sure that lone workers can keep in regular contact with you, especially those facing particular risks of accident or violence.
- Make sure you provide lone workers with any necessary training, instruction or demonstration.
- Keep confidential records – higher sickness rates or increased absenteeism levels may indicate potential problems.

Your aim is to ensure that lone workers are not at greater risk than other workers. If this is not possible, you should take appropriate action – you might decide that a particular individual is not suited to lone working, or that an activity is too dangerous to be carried out by one person on their own.

What is a safe system of work?

There is no legal definition of what constitutes a safe system of work and it will be a matter of 'fact and degree' for a court to determine.

Precedence was, however, set in the Court of Appeal in the 1940s when the then Master of the Rolls said:

> I do not venture to suggest a definition of what is meant by system. But it includes, or may include according to circumstances, such matters as physical lay-out of the job, the sequence in which work is to be carried out, the provision ... of warnings and notices and the issue of special instructions.
>
> A system may be adequate for the whole course of the job or it may have to be modified or improved to meet circumstances which arise: such modifications or improvements appear to me to equally fall under the heading of system.
>
> The safety of a system must be considered in relation to the particular circumstances of each particular job.

This means that a system of work must be tailored for each individual job.

What is the legal requirement for safe systems of work?

Section 2 of the Health and Safety at Work etc. Act 1974 sets out specifically that the employer is responsible for:

the provision and maintenance of plant and systems of work that are, so far as is reasonably practicable, safe and without risks to health.

The Confined Spaces Regulations 1997 also require employers to establish a safe system of work if work and entry into confined spaces cannot be avoided.

What are the provisions for a safe system of work?

Generally, developing a safe system of work will involve:

- carrying out a risk assessment
- identifying hazards and the steps that can be taken to eliminate them
- designing procedures and sequences that need to be taken to reduce exposure to the hazard
- considering whether certain things or actions need to be completed before others
- designing permit to work or permit to enter systems
- writing down the procedure
- training employees and others.

What is a permit to work?

A permit to work is a formal procedure, usually written or otherwise recorded, used to control certain types of work where there are specific hazards and risks associated with the job.

Increasingly, however, permit to work procedures are being implemented in many work environments so that written records exist to demonstrate that health and safety has been considered in the planning of the work.

Is there a specific format for permit to work systems?

No. There is no legally agreed format for a permit to follow – it can be designed in a variety of formats.

Are permit to work systems legally required?

While legislation does not specifically state that permit to work systems must be implemented, health and safety legislation requires that employers provide safe systems of work for their employees.

A safe system of work is a procedure that employees and others must follow so as to minimise the likelihood of accidents and incidents.

A permit to work system demonstrates that a safe system of work is being followed.

How does a permit to work system demonstrate that a safe system of work is being followed?

A good permit to work system or procedure sets out the hazards and risks associated with a work procedure and itemises the control systems or safe working practices that have been implemented so as to minimise the likelihood of accidents and incidents.

The permit to work will itemise the steps and checks that operatives need to take to ensure that they have considered hazards *before* they start the job.

Who is responsible for implementing a permit to work system?

The responsibility for implementing a permit to work system usually rests with:

■ the employer

or

■ the person in control of the works or building

or

■ the principal contractor.

Employers have a duty to ensure the health, safety and welfare of their employees and those who may be affected by their work activities.

Persons in control of the works or premises have legal duties to ensure the safety of those who resort to their premises and/or those who may be affected by the works (e.g. members of the public).

Principal contractors are responsible for a construction project where appointed under the Construction (Design and Management) Regulations 2015, and if just one contractor is appointed that contractor must set the standards for health and safety management. Clients can impose health and safety standards for a construction project, and many may require certain permit to work procedures to be adopted.

Who can issue permits to work?

Anyone who is competent in relation to health and safety and who understands the hazards and risks associated with the job can issue a permit to work.

Competency is the key issue. If people do not understand the hazards and risks associated with undertaking a task then they will not know how effective the control measures need to be to minimise the risk of injury and incident.

Effectively, the permit to work is the evidence that all hazards and risks have been considered before work starts, that suitable control measures have been considered before work starts and that suitable control measures have been put in place to ensure that the work progresses safely. If an incompetent person completes the permit then they may fail to consider the steps needed to address a specific hazard.

Permits can be issued by:

- maintenance managers
- site agents
- facilities management companies
- health and safety managers
- project managers
- specialist contractors.

Permits to work should not be issued by administration staff, secretaries, general employees, inexperienced operatives, etc.

What types of job should I consider as appropriate for a permit to work?

There are no hard and fast rules for what types of job suit a permit to work system, but some of the common work activities usually carried out under a permit to work procedure are:

- hot works
- working near or on live electrics
- roof access
- confined spaces
- working in harmful or hazardous atmospheres
- working in tanks or vessels
- working on or adjacent to fragile materials
- gas works
- railway works
- working in excavations
- working near asbestos materials
- working adjacent to water (e.g. rivers or canals).

Is a permit only relevant before work starts?

No. A permit to work is vitally important in identifying hazards and risks and the control measures needed to reduce risks before works start, but there needs to be a 'hand-back' procedure at the end of the job in order for the permit procedure to be worthwhile and effective.

What is a hand-back procedure?

Every permit needs to have a formal process for its issue and a formal process for its return.

As the employer or person in control of the premises, you will want to know that the works have been carried out safely and that the work area has been left safe.

A work permit may include a control measure whereby an alarm is turned off during the works so as to facilitate the work activity. It will be vital to ensure that the alarm has been reactivated at the end of the job so as to protect the people and the building. A 'hand-back' procedure for the permit will ensure that any procedures for reinstatement have been actioned.

Is it essential that people sign the permit?

Yes. The permit to work document is the written record that someone has taken responsibility for health and safety on the job. Whoever signs the permit is recording the fact that they have assessed the hazards and risks associated with the job and that they have agreed the safe system of work to be followed, or control measures to be implemented, so that persons carrying out the works or who may be affected by the works are kept safe.

Whoever signs the permit is taking on the responsibility to ensure that all operatives undertaking the job task receive information about the hazards and risks.

It is also important that the person who issues the permit signs it. They are confirming that they have reviewed the safe system of work and control measures that the contractor or person carrying out the work will follow, and that they are confident that all safety procedures have been considered and addressed.

If something were to go wrong and a prosecution ensue, it will be the formal signatures and the written procedures that would provide the defence that everything 'reasonably practicable' had been done to manage the risks.

What should a permit to work include?

A permit to work needs to include site-specific information that will address the hazards and risks associated with the job. All permits to work contain core information such as the:

- name of the contractor
- name of the supervisor or the person signing the permit
- address of the premises
- location of the work

- date the work is scheduled to occur
- duration of the work
- number of operatives proposed
- site-specific hazards and risks
- control measures to be implemented
- PPE to be used
- fire safety precautions
- hand-back procedures.

In addition, permits can contain sections to describe any training the operatives have had, evidence of risk assessments and method statements, and specific employer or client requirements.

The most important part of the permit is the section that describes the hazards associated with the job and the specific control measures that will need to be implemented.

PERMIT TO WORK

Site address: . . . .

Site agent: . . . .

Brief description of the work and location:
Sequence of work and control measures:
Supervision arrangements:
Individual responsible for controls and monitoring performance:
Plant and equipment to be used and operator training requirements:
Occupational health assessments (risk, noise, hazardous substances, etc.):
Measures to ensure the safety of third parties:
Environmental controls:
First aid and PPE requirements:
Emergency procedures:

Permit issued by: . . . .

Permit issued to: . . . .

Permit valid until: . . . .

LONE WORKER – RISK ASSESSMENT FORM

Name of employee undertaking lone work: . . . .

Department: . . . .

Title of Activity: . . . .

Location(s) of work: . . . .

Brief description of work: . . . .

Person fitness for work

Describe whether the employee is deemed medically fit to undertake the lone work activity (e.g. not on medication, not suffering from hypertension, not liable to panic attacks or not diabetic) – any medical condition must be considered and evaluated to ensure that the employee will not be put at any greater risk from the lone working activity.

Hazard identification

Identify all the hazards specific to the lone working activity; evaluate the risks (low/medium/high); describe all existing control measures and identify any further measures required.

Specific hazards should be assessed on a separate risk assessment form and cross-referenced with this document. Specific assessments are available for hazardous substances, stress, display screen equipment, manual handling operations, violence and aggression.

	Low	Medium	High
Workplace. Identify any hazard specific to the workplace/environment, which may create particular risks for lone workers (e.g. confined spaces)			
Process. Identify any hazards specific to the work process that may create particular risks for lone workers (e.g. electrical systems)			
Equipment. Identify any hazards specific to the work equipment that may create particular risks for lone workers (e.g. manual handling)			

	Low	Medium	High
Violence. Identify the potential risk of violence			
Individual. Identify any hazards specific to the individual that may create particular risks for lone workers (e.g. medical conditions, female, age, inexperience)			
Work pattern. Consider how the lone worker's work pattern integrates with those of other workers in terms of both time and geography			
Other: Please specify			

Continue on separate sheet, if necessary

Persons at risk
Identify all those who may be at risk.

Other employees			
Maintenance staff		Office staff	
Cleaning staff		Emergency staff	
Contractors		Visitors	
Others			

Training
Identify the level of information, instruction and training required. Consider the experience of workers.

Has necessary information, instruction and training been given?	
Expand and clarify, if necessary.	

Supervision

Identify the level of supervision required.

Is suitable supervision in place? (Identify all necessary supervisory measures)	
Periodic telephone contact with lone workers	
Periodic site visits to lone workers	
Regular contact (e.g. telephone, radio)	
Automatic warning devices (e.g. motion sensors)	
Manual warning devices (e.g. panic alarms)	
End of task/shift contact	
Other, specify	
Expand and clarify, if necessary	

Additional information

Identify any additional information relevant to the lone working activity, including emergency procedures and first aid provision.

Assessment carried out by:

Name: . . . .

Date: . . . .

Signature: . . . .

Review date: . . . .

Risk Assessments: Questions and Answers
ISBN 978-0-7277-6076-0

ICE Publishing: All rights reserved
http://dx.doi.org/10.1680/raqa.60760.187

Chapter 11
Persons with special needs

Persons with special needs include pregnant employees, young workers and employees with disabilities.

As an employer, what are some of the common risks to new or expectant mothers that I need to be aware of?

There are certain aspects of pregnancy that can be exacerbated by various work activities, and it is sensible to be aware of these so that special attention can be paid during the risk assessment process.

- Morning sickness and headaches:
  - consider early shift patterns
  - consider exposure to nauseating smells
  - consider exposure to excessive noise.
- Backache:
  - consider excessive standing
  - consider posture if sitting for prolonged periods
  - consider manual handling tasks.
- Varicose veins:
  - consider standing for prolonged periods
  - consider sitting positions and options for footrests.
- Haemorrhoids:
  - consider working in hot conditions.
- Frequent visits to the toilet:
  - consider the location of the work area in relation to toilets
  - consider the ease of leaving the workstation
  - consider whether software records absences
  - consider confined or restricted working space
  - consider ease of leaving the job quickly.
- Increasing size:
  - consider the use of protective clothing and uniforms
  - consider work in confined spaces
  - consider manual handling tasks.

- Tiredness:
  - consider whether overtime is necessary
  - consider shift patterns
  - consider whether evening work or early morning work is necessary
  - consider flexible working hours.
- Balance:
  - consider housekeeping to avoid obstacles
  - consider unobstructed work areas
  - specifically consider slippery floor surfaces
  - consider exposure to wet surfaces.
- Comfort:
  - consider too tight clothing
  - consider the temperature of the work area – too cold or too warm
  - consider 'overcrowding' with fellow employees
  - consider whether tasks need to be done at too great a speed.
- Stress:
  - consider anything that could cause a pregnant employee to become anxious about any working conditions.

What steps are involved in completing a risk assessment?

As with all risk assessment procedures, a planned approach is best.

Risk assessment is best broken down into steps or stages.

Risk assessment for new or expectant mothers is best started before any employee becomes pregnant – the employer is the best person to assess job activities that *could* cause problems for new or expectant mothers.

Initial risk assessment

Take into account any hazards or risks from your work activities (or those of others) that could affect employees of child-bearing age.

Risks include those to the unborn child or to a child of a woman who is breastfeeding.

Look for hazards that could affect all female employees, not just those who are pregnant.

Physical hazards:

- movement and posture
- manual handling

- noise
- shocks and vibrations
- radiation
- impact injuries
- using compressed-air tools
- working underground.

Biological hazards:

- working with micro-organisms that can cause infectious diseases.

Chemicals, gases and vapours:

- toxic chemicals
- mercury
- pesticides
- lead
- carbon monoxide
- medicines and drugs
- veterinary drugs.

Working conditions:

- stress
- working environment
- temperature
- ventilation
- travelling
- violence
- passive smoking
- working with display screen equipment
- working hours
- use of personal protective equipment
- use of work clothes and uniforms
- lone working
- working at heights
- mental and physical fatigue
- rest rooms, work breaks, etc.

Consider whether any of the above can harm any employee, but particularly new or expectant mothers.

The requirements to control many of the above hazards are contained in specific health and safety regulations (e.g. the Manual Handling Operations Regulations 1992).

Decide who is likely to be harmed, how, when and how often, etc. Remember that new or expectant mothers may be less tolerant to hazards than other workers, and so the degree of control needed to eliminate or reduce the risks may be greater than would normally be the case.

Consult your employees and inform them of any risks identified by the risk assessment. In particular, advise all female employees of child-bearing age of any risks that may affect them. Advise them of what steps are to be taken to reduce the risks.

What is a person-specific risk assessment and why is this necessary?

Generic risk assessments are an invaluable tool in helping to assess general hazards and risks in respect of pregnant employees or nursing mothers.

But everyone is different, and pregnancy affects women in different ways. So, not all pregnant women can be treated the same.

A person-specific risk assessment will ensure that the employer can demonstrate that they took due regard of the hazards and risks to the individual.

A person-specific risk assessment should be completed for each trimester of the woman's pregnancy, as the effects of hazards and risks will vary – something that was acceptable in the first 3 months may not be quite so acceptable when the employee is 6 months pregnant.

What needs to be considered in the risk assessment for young workers?

First, the risk assessment must be completed *before* the young person starts work or work experience.

Each young person must be told what the hazards and risks are and must have control measures explained, etc.

The risk assessment must:

- consider the fact that young people are inexperienced in work environments
- consider that young people are physically and mentally immature
- consider that young people lack knowledge in work procedures

- consider that young people are inexperienced in perceiving danger
- consider that young people's literacy skills may be less than ideal
- consider all the control measures necessary to reduce or eliminate the hazard
- consider that personal protective equipment, etc., if identified as a control measure, may be sized for adults and may not therefore be 'suitable and sufficient'
- be kept up to date
- be relayed to and discussed with the young person
- identify training needs for the young person
- consider the tools and equipment the young person will use – whether there is an age restriction (e.g. on dangerous machines)
- consider the layout of the workplace
- consider the environmental hazards
- consider any hazardous substances in use.

Are there any risks that young people cannot legally be exposed to?

People under the age of 18 years must not be allowed to do work which:

- cannot be adapted to meet any physical or mental limitations they may have
- exposes them to substances that are toxic or cause cancer
- exposes them to radiation
- involves extreme heat, noise or vibration.

However, if people are over the age of 16 years they may be able to undertake or be exposed to the above tasks and hazards if it is necessary for their training and if they are under constant supervision by a competent person.

Children below the school leaving age must *never* be allowed to undertake the tasks or be exposed to the hazards listed above.

What training and supervision do young people require under health and safety?

Young people need training when they start work – *before* they undertake any work activity, process or task. They must be trained to do the work without putting themselves or others at risk.

It is important to check that young people have understood the training and information they have been given:

- Do they understand the hazards and risks of the workplace?

- Do they understand the basic emergency procedures (e.g. fire evacuation, first aid and accident reporting)?
- Do they understand the control measures in place to eliminate or reduce risks?
- Do they understand their responsibilities as employees not to interfere with safety equipment, not to fool about, etc.?

Young people must be regularly supervised by competent people (i.e. those who understand that a young person may inadvertently put themself or others at risk because they do not know any better or cannot perceive the risk).

Are specific risk assessments required for employees with disabilities?

Employers have duties under various regulations to ensure the safety of all employees, including those with disabilities. Where employees with disabilities carry out work-related tasks that could expose them to hazard and risk, those tasks will need to be assessed.

The Management of Health and Safety at Work Regulations 1999 require employers to consider the capability of employees to carry out a task, and therefore the employer will need to assess the individual capability of the employee to determine whether their disability creates additional risks to their safety or health or that of others.

How does an employer assess whether a person's disability will put them or others at risk?

An employer must not make assumptions about an employee's condition and should consider a number of things:

- Have the risks to all employees been assessed properly and appropriate control measures put in place?
- It may be that appropriate changes to the work equipment and environment could significantly reduce the risk and take the issue of disability out of the equation.
- Does the employee's condition create an increased risk to their health and safety or the health and safety of others?
- The condition might be well managed and the employer may conclude that review of the situation at regular intervals is sufficient.
- If so, can these risks be prevented or adequately controlled through normal health and safety management?
- Can the risks be addressed by allowing other colleagues to do certain elements of the activity or by providing suitable, alternative equipment (e.g. automated equipment to reduce manual handling) or by changing systems of work?

If not, what reasonable adjustments could be put in place to prevent or adequately control the residual risks?

The employer might be able to apply for financial assistance through the Government's Access to Work scheme to cover the cost of new equipment.

Be sure to consult with the employee themselves and their colleagues, seeking opinions and ensuring they are involved in discussions that affect them. Those involved in the work often propose good solutions.

The Equality Act 2010 requires employers and service providers to make reasonable adjustments for people with disabilities. What does this mean?

Equality law recognises that bringing about equality for disabled people may mean changing the way in which employment is structured, the removal of physical barriers and/or providing extra support for a disabled worker.

This is the duty to make reasonable adjustments.

The duty to make reasonable adjustments aims to make sure that, as far as is reasonable, a disabled worker has the same access to everything that is involved in doing and keeping a job as a non-disabled person.

When the duty arises, you are under a positive and proactive duty to take steps to remove, reduce or prevent the obstacles a disabled worker or job applicant faces.

You only have to make adjustments where you are aware – or should reasonably be aware – that a worker has a disability.

Many of the adjustments you can make will not be particularly expensive, and you are not required to do more than what is reasonable for you to do. What is reasonable for you to do depends, among other factors, on the size and nature of your organisation.

If, however, you do nothing, and a disabled worker can show that there were barriers you should have identified and reasonable adjustments you could have made, they can bring a claim against you in an employment tribunal, and you may be ordered to pay compensation as well as make the reasonable adjustments.

In particular, the need to make adjustments for an individual worker:

■ must not be a reason not to promote a worker if they are the best person for the job with the adjustments in place

- must not be a reason to dismiss a worker
- must be considered in relation to every aspect of a worker's job

provided the adjustments are reasonable for you to make.

PREGNANT WOMEN – EXAMPLE RISK ASSESSMENT – GENERIC RISKS

Task/activity Working while pregnant, breastfeeding or as a new mother – general office work, lifting loads, sitting at desk and driving
Who is at risk from the activity? Pregnant and breastfeeding women and those that have given birth in the last 6 months (new mothers)

What are the hazards (dangers)?

Excessive strain	Overstretching/twisting
Heat	Tiredness
Chemical	Slips/trips and falls
Lifting	Display screen equipment/workstations

What are the potential outcomes from the hazards?

Fainting	Misjudgement
Chemical exposure	Slips/falls
Back strains	Personal injury
Damage to unborn baby	Repetitive movement
Eyestrain/headaches	

What is the likelihood of the risk occurring?

High

Medium

Low

How do we currently control these risks?
- Allow regular breaks to relieve muscle strain and tiredness, which may add to misjudgement.
- Ensure the environment is well ventilated, drinking water provided and regular breaks are taken.
- Do not lift excessive weights. Always get help with lifting. Ensure training is given.
- Read control of substances hazardous to health data prior to chemical use.
- Wear appropriate protective equipment for chemicals used.
- Minimise activity in order to avoid stress, excessive stretching, twisting and strain to the body.
- Keep floor surfaces slip/trip hazard free and in good condition.
- Ensure the line manager undertakes a review of the pregnant woman and her activities by using the form in health and safety policy to address site-specific and person-specific duties and actions.

- Take regular breaks from driving – request a back support for the car if necessary.
- Avoid undertaking long journeys where regular breaks are not possible.
- Take regular breaks from display screen equipment (DSE) – 10 minutes in every hour undertaking different activities.
- Provide a footrest at the DSE if necessary.
- Ensure a DSE self-assessment is undertaken (refer to the health and safety policy)
- Ensure the availability of a non-smoking rest area, with seating.

References

Health and safety policy

Assessment form to support pregnancy risk assessment

What else can we do/what else is required?
- Monitor activities of employee.
- Do not work if feeling unwell – advise manager of this.
- Do not continue to work if unable (i.e. near to birth).

Who prepared the risk assessment and when?

Who needs to know about these findings?

Managing director

Managers

All staff (all women of child-bearing age)

PREGNANT WOMEN – RISK ASSESSMENT FORM – ADDITIONAL PERSON-SPECIFIC INFORMATION

Employee name: . . . .

Expectant/new mother *(Please highlight applicable category)*

Premises name: . . . .

Please indicate stage of pregnancy: 1–3 months   4–6 months   7–9 months

New mother   Breastfeeding mother

NB: This form is to be completed for all stages as indicated above.

In addition to the general pregnancy risk assessment available, this specific pregnancy assessment reviews the working conditions, environment and medical status of the pregnant employee through each stage of the pregnancy and the employee's return to work after the birth.

The line manager must complete this form *with* the pregnant employee at each of the above stages. The guidance contained within the assessment form will give recommendations and indications of action to be taken where hazards may be identified.

Manual handling
If the answer is 'yes' to any of the tasks/activities detailed below, give exact details of tasks under 'detail'. Also if the answer is 'yes', where 'action to be taken' details 'minimise task', consider with the employee what is reasonable, what they are capable and what they are comfortable with. They may usually be able to lift and carry a limited amount of weight. Also, movements such as 'reaching upwards' or 'twisting' will need to be reviewed but so long as they are not excessive, and do not involve, for example, significant risk of injury or lifting weight while doing so, this sort of activity will normally be reasonable to expect.

Is there risk of manual handling injury from:	Yes	No	Detail (if yes)	Action to be taken (if yes)
Lifting filled crates/other heavy goods				Eliminate task
Carrying food deliveries				Minimise task
Carrying stock				Minimise task
Moving furniture				Minimise task

Pushing/pulling items				Minimise task
Reaching upwards				Minimise task
Twisting movements				Minimise activity where practical
Is there risk of manual handling injury from:	**Yes**	**No**	**Detail (if yes)**	**Action to be taken (if yes)**

Working environment

Consider whether any exposure to the factors below would mean an increased risk of injury. For example, if working behind the bar the risk of slips may be elevated, but consider measures to help prevent and deal with spillages. Are they adequate, do they need improvement? Detail any actions to be taken as a result of your assessment.

Is there any risk of injury from:	**Yes**	**No**	**Detail (if yes)**	**Action to be taken (if yes)**
Slips, trips and falls *(Consider trip hazards, risks of falling)*				
Cramped working space *(Consider any confined spaces or cramped working areas that the pregnant employee may have to enter – e.g. an area where they have to bend down/slouch to work)*				
Excessive cold *(Consider the walk-in freezer, for example – will the employee have to be in there for any significant period of time, and if so what can be done? Can they limit the time spent in areas such as a freezer; can they wear protective, warmer clothing?)*				
Excessive heat *(Consider areas such as the kitchen – can regular breaks be taken, is drinking water available, is the ventilation working?)*				

	Yes	No	Detail	Action
Excessive travelling distances *(Does the employee have to walk very far from work areas? Does she have to travel far to work by public transport? Contact the personnel department – can the employee be transferred to another premises for a temporary period, or can she stay at a local hotel if she finishes a shift late at night, e.g. past 11 p.m.?)*				
Continual use of stairs *(Does the employee have to go up and down stairs at work? Is this excessive and likely to lead to increased fatigue? Can it be avoided? What can be done?)*				
Exposure to excessive noise *(If the premises has very loud music at a time when the pregnant employee is working, is this excessive? Is the ventilation working in the premises containing other mechanical equipment without too much noise? Detail any action needed.)*				

Working time/activity

Is there a risk of injury, ill health or fatigue from:	Yes	No	Detail (if yes)	Action to be taken (if yes)
Standing for long periods (i.e. over 2 hours without breaks)				Provide a seat/stool to sit down at quieter periods
Working excessive shifts (i.e. not allowing an 11-hour break in 24-hour shift work)				Contact the human resources department to reduce shifts
Working more than 48 hours				Reduce hours to below 48 hours per week. Contact the human resources department
Lack of seating				Provide seated, non-smoking rest area for breaks

Lack of breaks (should be at least 40 minutes in every 6 hours)				Provide additional breaks. Contact the human resources department

Workstations/VDU stations
Review the use of any VDU by the pregnant employee – refer to the relevant section in the health and safety policy. Detail any necessary action to increase comfort if needed, and allow regular breaks away from the screen (at least 10 minutes in every hour).

Is there risk of injury, ill health or fatigue from:	Yes	No	Detail (if yes)	Action to be taken (if yes)
Long periods at the workstation (i.e. over 2 hours at any one time)				Increase breaks away from the screen to 10 minutes in every hour doing different tasks
Glare from screens				Fit a screen glare guard
Poor work position				Improve seating, etc.
Inadequate space (should be enough space to sit and work comfortably)				Contact the human resources department
Lack of breaks				Contact the human resources department

Use of chemicals, etc.
Check that all chemicals used are those approved in the COSHH manual. Review the employee's contact with body-fluids spillages – can they be avoided to an extent (i.e. can the employee avoid clearing body fluid spillages, etc.)?

Is there any exposure to:	Yes	No	Detail (if yes)	Actions to be taken (if yes)
Chemicals/substances not in COSHH manual				Contact the human resources department
Fumes, excessive smells, dust, etc.				Contact the human resources department
Biological agents (e.g. body fluids)				Avoid clearing up vomit, urine, blood, etc. using spill packs

Night working

If the employee is working during the evening/night shifts (e.g. past 8 p.m.), consider what can be done to give extra breaks/rest. Can the employee sit down during quieter periods? Is there access to a seated, non-smoking rest area? How will she get home after a night shift? Consider options as detailed in the 'Working environment' above.

	Yes	No	Detail (if yes)	Actions to be taken (if yes)
Will the employee be working at night (i.e. on a night shift (past 8 p.m.))?				
Will the employee be working until 11.00 p.m., 12.00 p.m. or 1.00 a.m.?				

Wherever 'yes' has been indicated above, make sure sufficient detail has been recorded. If you need to provide further detail, include it in this section.

Other issues

Describe any other issue that may affect the individual's health and safety due to being either an expectant or new mother, or because of an existing medical condition. For example, pre-existing medical conditions that may affect a woman during pregnancy or afterwards include:

- *diabetes*
- *any heart condition*
- *joint/bone conditions (e.g. osteoporosis, previous injuries causing pain/discomfort)*
- *previous miscarriage.*

THE PREGNANT EMPLOYEE IS RESPONSIBLE FOR CONTACTING THE HUMAN RESOURCES DEPARTMENT IF SHE FEELS THAT THERE MAY BE A MEDICAL CONDITION THAT SHOULD BE RECORDED WHICH SHE HAS NOT RAISED DURING THE ASSESSMENT.

Signed (employee): . . . .

Date: . . . .

Signed (line manager): . . . .

Date: . . . .

Retain a copy for the personnel file and copy to the human resources department.

EMPLOYEES WITH DISABILITIES – RISK ASSESSMENT FORM

Name of employee: . . . .

Name of employer: . . . .

Location of employment: . . . .

Job title: . . . .

Job tasks: . . . .

. . . .

Nature of disability as it affects health and safety: . . . .

. . . .

. . . .

Hazards associated with the tasks: . . . .

. . . .

Additional hazards created due to the nature of the disability: . . . .

. . . .

. . . .

Control measures implemented: . . . .

. . . .

. . . .

Additional control measures in order to ensure the safety of the employee with a disability:

. . . .

. . . .

Additional information: . . . .

Agreement with disabled employee: . . . .

Date of assessment: . . . .

Carried out by: . . . .

Date: . . . .

Risk Assessments: Questions and Answers
ISBN 978-0-7277-6076-0

ICE Publishing: All rights reserved
http://dx.doi.org/10.1680/raqa.60760.203

Chapter 12
Personal protective equipment

What is personal protective equipment (PPE)?

The Personal Protective Equipment at Work Regulations 1992 define PPE as:

> all equipment (including clothing affording protection against the weather) which is intended to be worn or held by a person at work and which protects him from one or more risks to his health or safety.

PPE therefore includes, but is not limited to:

- safety helmets
- ear protection
- eye protection
- gloves
- safety shoes and boots
- high-visibility jackets
- respiratory masks
- breathing apparatus.

Ordinary working clothes (e.g. uniforms) are exempt from the regulations unless they are worn for health or safety reasons (e.g. steel chain sleeved jackets).

What does the law require an employer to do in respect of PPE?

An employer has to carry out a risk assessment of the job tasks and activities that employees have to undertake, and must determine the control measures necessary to reduce hazards and risks to the lowest possible level.

If elimination of the hazard at source, engineering controls or a safe system of work will not reduce the hazard to an acceptable level, the employer must provide employees with appropriate PPE.

PPE provided to employees must be free of charge.

Regulation 6 of the Personal Protective Equipment at Work Regulations 1992 requires an employer to ensure that an assessment is made in respect of PPE needs. This must include:

- identifying risks not avoided by other means
- defining the characteristics required of the PPE
- comparing these with the characteristics of the PPE available.

In all but the simplest of cases the assessment should be recorded and kept readily available.

Will the risk assessment carried out under the Management of Health and Safety at Work Regulations 1999 be sufficient to comply with the Personal Protective Equipment at Work Regulations 1992?

Yes, provided the risk assessment has given due regard to the need for PPE and has determined whether the PPE is suitable.

The risk assessment under the Management of Health and Safety at Work Regulations 1999 will determine what health and safety measures are needed to comply with legal requirements.

If the conclusion is that PPE is required, another risk assessment is required under the Personal Protective Equipment at Work Regulations 1992 in order to determine if the PPE is suitable.

In reality, the two risk assessments are combined.

Remember that a risk assessment under the management regulations must be 'suitable and sufficient'. There is no such qualification for assessments under the PPE regulations.

When does a risk assessment under the Personal Protective Equipment at Work Regulations 1992 have to be carried out?

An assessment has to be made before choosing any PPE that has to be provided.

What are the key requirements for PPE?

Many factors need to be considered when choosing PPE because choosing the wrong sort, an ill-fitting type or an inappropriate specification may cause the employee to be exposed to greater hazards than having no PPE at all.

Often, people believe themselves to be adequately protected because they trust their PPE. Sometimes such trust is misplaced because the equipment is not providing protection.

In general, PPE must:

- be capable of adjustment to fit correctly
- be appropriate for the risks and conditions
- take account of ergonomic requirements and the state of the wearer's health
- prevent or adequately control the risks to which the wearer is exposed, without adding to those risks
- comply with appropriate product safety and other standards (e.g. CE mark)
- be compatible with any other types of PPE being worn.

Who is responsible for monitoring and cleaning PPE?

The employer has a duty to maintain all PPE provided for his employees. PPE must be maintained in an efficient sate, in efficient working order and in good repair.

Specific arrangements may be needed for:

- inspecting PPE
- maintaining PPE
- cleaning PPE
- disinfecting PPE
- replacing PPE
- examining PPE
- testing PPE.

The company safety policy should state who is responsible for carrying out proactive PPE inspection and testing procedures.

Defective and inappropriate PPE is often a greater hazard than no PPE.

It is sometimes appropriate to give 'users' some responsibility to check their own PPE (e.g. to ensure that face masks fit correctly). Employees who use PPE must know what to do to report that their PPE may be defective.

What records need to be kept?

Where the employer has five or more employees a record of any risk assessment should be in writing, but it is good practice to keep a record no matter how many employees there are in the company.

It is essential to keep a record of the assessment, including why you chose the PPE you did, what its purpose is, when it was issued and to whom, what training have employees had and when the PPE should be inspected and replaced.

Most employers keep an employee record sheet so that they know who has what and what to expect to be returned to them if an employee leaves the business.

What training does an employee have to have in the use of PPE?

Regulation 9 of the Personal Protective Equipment at Work Regulations 1992 requires employers to give employees adequate and appropriate information, instruction and training on any PPE they are expected to wear in order to safeguard them from hazards at work.

The training, information or instruction must include:

- why and when PPE must be used
- how to use it
- its limitations
- arrangements for its maintenance and/or replacement.

'How to use it' training may involve a practical demonstration of how to wear masks, breathing apparatus, goggles, etc.

Records of training must be kept. It is not sufficient merely to give employees PPE to wear or use. They *must* be given information on why to use it, how to wear it, etc.

Case study

Mrs W worked for the county council as a cleaner. She was given rubber gloves to wear but did not really know when to wear them, why she needed to, etc. So she and her colleagues rarely wore them.

Mrs W used a range of detergents and cleaning chemicals over the years of her employment, and after some years she started to develop eczema on her hands and wrists. Dermatitis was diagnosed, and Mrs W's GP advised her to wear cotton gloves under her rubber gloves. She did this but the eczema got worse and spread to her face and other parts of her body. She gave up work on medical advice.

Mrs W sued her employer for damages on the grounds that it had not protected her health and safety.

The court found in favour of Mrs W and awarded her compensation.

The court decided that it was not enough for the county council to gives its employees PPE (i.e. gloves). In order to discharge its duty as an employer has to provide a safe system of work, and the council should have warned the cleaners about the dangers of handling the cleaning chemicals without protection and should have instructed them to wear gloves at all times.

The judge also concluded that the risks of developing dermatitis from using cleaning chemicals were well known, and the employer ought to have known of the risks. But the risks were not so well known that they would have been obvious to the staff without them receiving any necessary warning or instruction.

What are the responsibilities of employees to wear PPE given to them by their employer?

Employees have duties under the Personal Protective Equipment at Work Regulations 1992 to:

- use PPE in accordance with their training and instruction
- return PPE after use to the place specified by the employer
- immediately report loss of or damage to the PPE they use.

If an employee has had appropriate training, information and instruction and yet persists in abusing or mis-wearing the PPE provided, and has an accident because the PPE did not protect them, they may be considered contributorily negligent with regard to their injuries and therefore not eligible for compensation, and/or they may be guilty of a criminal offence under Section 7 of the Health and Safety at Work etc. Act 1974 (which requires employees to cooperate with their employers and not to jeopardise their or others' safety).

As an employer, can I force my employees to wear the PPE I have provided?

Employers have a duty under Regulation 10 of the Personal Protective Equipment at Work Regulations 1992 to take all reasonable steps to ensure that PPE is properly used by employees.

Employers are therefore expected to take a proactive approach to enforcing their PPE rules and to introduce appropriate measures to ensure that the message is understood.

Many employers include the flaunting of PPE requirements as gross misconduct under employment contracts and deal with breaches as disciplinary issues.

What factors need to be considered when choosing PPE?

There are three distinct areas to consider when choosing PPE:

- the workplace
- the work environment
- the wearer of the PPE.

The workplace
- What are the hazards to be controlled?
- What machinery is in use?
- How many people work in the area?
- What are the known risks?
- What is the PPE expected to do?
- Will people need to be able to move freely when using the PPE, will they need dexterity to use equipment, etc.?

The work environment
- Will excessive temperatures be generated?
- Will ventilation be available?
- Is the PPE to be used in a confined space?
- Could the PPE cause other hazards (e.g. prevent alarm bells from being heard)?

The wearer of the PPE
- Fit – what size is needed, does it feel comfortable to wear, can it be adapted to the wearer (adjustable straps, etc.), does any physical feature affect its fit (e.g. beards and face masks)?
- Training – do users know what they have to wear and why, what the hazards and risks are, how to wear the PPE, how to inspect it before use, how to report defects, etc.?
- Acceptability and comfort – will users wear the PPE for prolonged periods, will it be too heavy or cumbersome, will it slow users down so that 'piece work' may be affected, does the user have a choice in particular types or style of PPE?
- Interference – will the PPE work with other PPE (e.g. goggles and masks), will it prevent other controls being used (e.g. ear defenders will prevent warning buzzers, etc. from being heard), will goggles restrict vision, etc.?
- Management commitment – do employees see a culture of 'do as I do' and not 'do as I say but don't do'? Are management, including supervisors, leading by example?

What sort of PPE is necessary for different job tasks and activities?
Hearing protection
There are two main types of hearing protection: things placed in the ear canal (ear plugs) and objects placed around the outer ear (ear muffs).

Ear plugs:

- fit inside the ear canal and impede the passage of sound energy to the ear drum
- are made from glass down, rubber or foam

- are usually disposable
- allow some sound to reach the ear drum
- create hygiene problems if reused
- are unlikely to produce a good fit
- are available in various sizes but could be too small or big for an individual ear.

Ear muffs:

- are rigid cups that fit over the outer ear and are held in place by a head band
- are unlikely to give a good fit
- allow some sound to reach the ear drum
- are difficult to wear with glasses, goggles, etc.

Respiratory protective equipment

There are two broad categories of respiratory protective equipment: respirators, which purity the air to be breathed in, and masks, which filter out particles.

Breathing apparatus purifies the air and draws uncontaminated air from an independent source to the wearer (e.g. from air oxygen tanks).

Face masks can be full-face or half-face masks.

Dust masks eliminate particulate matter from being breathed in by containing it in the filtering medium.

Eye protection

Eyes can be protected by goggles, spectacles or face masks.

Spectacles/glasses:

- are used for protection against low-energy projectiles
- do not protect against dust
- are easily displaced
- are usually incompatible with other PPE.

Goggles:

- are used for protection against high-energy projectiles
- protect against dust, etc.
- protect the whole of the eye and the surrounding face area
- may mist up and impede vision
- may be uncomfortable and cumbersome
- protect against splashes.

Full-face masks:

- provide full-face protection
- may be uncomfortable.

Head protection
There are two general types of head protection: safety helmets and bump caps.

Safety helmets:

- are worn on construction sites
- protect against impact damage
- deteriorate in sunlight.

Bump caps:

- provide less protection from impact damage than safety helmets
- are easier to wear than safety helmets, especially in confined spaces.

Hand/arm protection
Gloves and gauntlets:

- may be made of leather, which provides good protection
- may be made of chain mail, which protects against cuts
- may be made of rubber, PVC, nylon, cotton, latex or cloth
- may cause dermatitis
- are resistant to fluids
- may corrode in some substances
- are easily damaged
- may be cumbersome
- may prevent good gripping
- are uncomfortable to wear.

Safety footwear
Safety footwear is usually used where objects may fall onto the feet, where objects may pierce the soles of ordinary footwear, in wet conditions, etc.

Safety boots/shoes:

- have steel toe caps
- have steel soles

- are lightweight
- are cumbersome
- protect against oils, chemicals, etc.

Wellingtons:

- have steel toe caps
- protect against water and fluids
- may not provide a good grip.

Skin barrier creams

Skin barrier creams are used to assist in protecting the hands from direct contact with fluids, cleaning chemicals, etc.

Other types of PPE

- safety harnesses
- lanyards.

The eight requirements of the Personal Protective Equipment at Work Regulations 1992

- PPE is to be provided as the last resort.
- A test of the suitability of the PPE is to be applied.
- There must be compatibility between different types of PPE.
- PPE maintenance and replacement schemes must be in place.
- Places to keep PPE must be provided.
- Proper use must be made of the PPE.
- Users must report defects in or loss of PPE.
- Information, instruction and training in the need for and the use of PPE must be provided.

PERSONAL PROTECTIVE EQUIPMENT – RISK ASSESSMENT FORM

Name of company: . . . .

Assessment carried out by: . . . .

Job activity/type under review: . . . .

Name of operatives undertaking tasks: . . . .

Hazards associated with task/activity: . . . .

. . . .

Personal protective equipment required

Type	Purpose	When to be worn/used

Maintenance, inspection, cleaning procedure
Training, information and instruction
Review of assessment (date)

PERSONAL PROTECTIVE EQUIPMENT – RECORD SHEET

Name of employee: . . . .

Job occupation/title: . . . .

Main hazards associated with job tasks/activities: . . . .

. . . .

Personal protective equipment issued

Type	Condition	Date issued

Training information and instruction undertaken

Subject	Date	Signed

Signed (employee): . . . .

Date: . . . .

Signed (supervisor): . . . .

Date: . . . .

Risk Assessments: Questions and Answers
ISBN 978-0-7277-6076-0

ICE Publishing: All rights reserved
http://dx.doi.org/10.1680/raqa.60760.215

Chapter 13
First aid

What is the main piece of legislation that covers first aid at work?

The Health and Safety (First Aid) Regulations 1981 set the standards for first aid at work.

The main scope of the regulations is as follows:

Every employer must provide first-aid equipment and facilities that are adequate and appropriate in the circumstances for administering first aid to employees.

Employers must make an assessment to determine the first-aid needs posed by the workplace. First aid precautions will depend on the type of work being carried out, and therefore the risk involved.

Employers should consider the need for first-aid rooms, the first-aid needs when employees working away from the premises, the needs when the employees of more than one employer are working together, and the needs for non-employees.

Once an assessment has been made the employer can work out the number of first-aid kits necessary by referring to the Health and Safety (First-Aid) Regulations 1981. Guidance on Regulation (L74, 2013).

Employers must ensure that adequate numbers of 'suitable persons' are provided to administer first aid. A 'suitable person' is someone trained to an appropriate standard in first aid.

In appropriate circumstances the employer can appoint an 'appointed person' instead of a first aider. This person will take charge of any medical situation (e.g. call an ambulance) and should be able to administer emergency first aid.

Employers must inform all employees of their first-aid arrangements and identify trained personnel.

What is first aid at work?

First aid at work covers the arrangements that an employer must make to provide employees with adequate first aid attention while they are at work.

Employees may suffer injury or ill health while at work. This may be due to a work activity or the work environment, or employees may become ill for other reasons while at work.

In serious cases immediate emergency attention needs to be provided and an ambulance must be called.

First aid at work is designed to save lives and to prevent minor injuries or incidents escalating into serious ones.

As an employer, what do I need to do?

The Health and Safety (First-Aid) Regulations 1981 require employers to provide adequate and appropriate equipment, facilities and personnel to enable first aid to be given to employees if they are injured or become ill at work.

The regulations set out some minimum first-aid provisions to be provided on an employer's site as follows:

- a suitably stocked first-aid kit
- an appointed person to take charge of any incident and the first-aid arrangements
- first-aid facilities to be available at all times when people are at work.

The key phrases regarding first aid are: 'adequate and appropriate' and 'suitable and sufficient'.

How the two requirements will be satisfied depends on the specific circumstances of the workplace.

In order to ascertain what appropriate first aid is needed, employers will need to carry out a risk assessment of first-aid needs.

What should an employer consider when assessing first-aid needs?

The risk assessment process requires employers to consider the hazards and associated risks involved in the work activities.

Small businesses will need only the simplest of risk assessments and basic first-aid provision.

Larger businesses will need to consider the following:

- *Step 1*. Consider what risks of injury and ill health are associated with the work practices and activities.
- *Step 2*. Are there any specific risks that can be clearly identified? For example:
  - working with hazardous chemicals
  - working with dangerous tools
  - working with dangerous machinery
  - working with dangerous loads
  - working with animals.
- *Step 3*. Are there areas within the business where risks may be greater because of the environment, and might these need additional first-aid facilities? For example:
  - research laboratories
  - pathology laboratories
  - hot working environments
  - cold working environments.
- *Step 4*. Consider the businesses record of accidents and injuries. What types have occurred? How serious have they been? Is there any evidence which shows that extra precautions, etc. are necessary? Where, when and why did the incidents happen?
- *Step 5*. How many people are employed on the site? How many are permanent, temporary, etc.? How familiar are they with procedures, processes, etc.? Are there any young or inexperienced workers who are more likely to have an accident or suffer ill health than the rest of the workforce? Does anything extra need to be done for people with disabilities?
- *Step 6*. What and how are the buildings used for, are they spread out, do employees work out of doors, can they access all parts of the building? Where might first-aid provisions and facilities be located so that they are available to all?
- *Step 7*. Do employees work outside of normal working hours, work overtime, work alone, etc.? If so, how might they raise an alarm? Do employees travel frequently?
- *Step 8*. Would emergency services be easy to summon? Could these services gain access to the site or building out of normal working hours? Are the workplaces accessible?

What are some of the steps an employer needs to take in order to make suitable provision for first aid?

The assessment of the likelihood and frequency of injury and ill health will determine what needs to be done to ensure that suitable first-aid provision (i.e. emergency aid) is in place.

First-aid kits need to be provided – should there be one in a central location, or will it be preferable to have several smaller ones easily accessible to employees?

An appointed person or a trained first aider will need to be appointed depending on the number of employees and the severity of risk of injury, etc.

A first-aid room may be required, depending on the number of employees.

Emergency procedures for calling the emergency services will be needed.

Special consideration will be needed for people with disabilities.

The type of first-aid equipment required will need to be considered. For example, will eyewash stations be needed?

Workers from other companies who are working in the premises will need to be considered – who is providing first-aid facilities for them (e.g. on construction sites)?

Who is an 'appointed person'?

An appointed person is someone who is not necessarily trained in emergency first-aid treatment but who is appointed to take charge of an incident and call the emergency services if required.

The appointed person usually also looks after the first-aid equipment and ensures that it is adequately stocked, in the right location, etc.

Employers who have 50 or fewer employees and whose business falls in the category of 'low risk' (i.e. office environments, libraries, retail shops, etc.) need only appoint an 'appointed person'.

There should also be 'reserve' appointed persons to cover for holidays and sickness.

What is a first aider?

A first aider is someone who has undergone training in first aid and holds a current certificate in first aid at work. All first-aider training courses must be approved by the Health and Safety Executive (HSE), so ask training providers for details of their HSE registration documentation, etc.

Employers who have more than 50 employees usually need to appoint a trained first aider.

First aid training courses last for 4 days and the certificate is valid for 3 years. After that, retraining is required.

A trained first aider can administer first aid, with the primary purpose being to prevent injuries from getting worse rather than to try to treat people or provide medical expertise.

How many first aiders or appointed persons does an employer need?

The regulations do not set down a hard and fast rule on the number of people to be appointed. The number really depends on the type of work activity and the likelihood of injury.

The law requires an employer to make an assessment of the number of people required. As long as the number of appointed people can be explained and justified as being suitable and sufficient, the law will be satisfied.

However, the Health and Safety (First-Aid) Regulations 1981. Guidance on Regulation (L74, 2013) gives information on suitable numbers of appointed persons and first aiders.

Enough people should be nominated and trained so that absences can be covered.

A suggested ratio of appointed persons or first aiders is as follows:

Shops, offices, libraries	<50 employees	1 appointed person
	50–100 employees	1 first aider
	>100 employees	1 first aider plus an extra person for every 100 employees
Food processing, warehouses	<20 employees	1 appointed person
	20–100 employees	1 first aider for every 50 employees
	>100 employees	2 first aiders plus an extra person for every 100 employees
Construction sites, industrial sites, manufacturing, spray shops, chemical industries, etc.	<5 employees	1 appointed person
	5–50 employees	At least 1 first aider
	>50 employees	1 first aider for every 50 employees

What should be in a first-aid kit?

There is no legal list of items that should be in a first-aid box, although there is guidance in the Health and Safety (First-Aid) Regulations 1981. Guidance on Regulation (L74, 2013).

The contents of a first-aid kit really depend on the risk assessment for first-aid facilities, which every employer must complete.

First-aid kits should *not*, however, contain any medicines (e.g. aspirin or paracetamol) because no one is trained to issue medicines and the recipient could be allergic to a substance given.

The contents of a first-aid kit are relatively straightforward, and the following would be sensible contents for a low-risk work environment:

- 20 individual sterile adhesive dressings (plasters) of varying sizes
- two sterile eye pads
- four sterile triangular bandages
- safety pins
- six medium-sized sterile wound dressings
- two large sterile wound dressings
- disposable gloves
- an advice leaflet.

All dressings, plasters, etc. should be individually wrapped. Dressings have a 'shelf-life', and dates should be checked and anything that is out of date replaced because it may be no longer sterile, etc.

Is an employer responsible for providing first-aid facilities for members of the public, customers, etc.?

No, the law on first-aid applies to employees while they are at work.

However, it is good practice to consider the needs of customers or the public when completing the risk assessment. The HSE strongly recommends that the public and customers are included in first-aid provision.

Are there any other responsibilities that an employer has in respect of first aid?

It is a legal requirement for employers to inform their employees of the first-aid arrangements that are in place.

Notices can be displayed advising where the first-aid kit is located and who the appointed persons or first aiders are.

Special arrangements will need to be considered for any employees with language problems, learning disabilities, etc.

When is it necessary to provide a first-aid room?

There is no legal requirement within the Health and Safety (First-Aid) Regulations 1981 for employers to provide a first-aid room. As with the provision of any first-aid facilities, the need is determined in the risk assessment.

Guidance in the Health and Safety (First-Aid) Regulations 1981. Guidance on Regulation (L74, 2013) indicates that it would be good practice to provide a first-aid room when there are 150 employees or more.

If a first-aid room is provided it must:

- be easily accessible to all employees
- be easily accessible for emergency services, people carrying stretchers, etc.
- have heating, lighting and ventilation
- have hand-washing facilities, preferably a sink
- have drinking water
- contain a chair, couch, table/desk, etc.
- contain first-aid materials (first-aid kits, etc.)
- contain a refuse bin
- contain blankets, pillows, etc.
- have some method of raising the alarm
- contain record books
- have surfaces that are easily cleaned and disinfected
- be clean and tidy
- contain suitable first-aid information (e.g. names of first aiders).

A first-aid room should be used exclusively as a first-aid room so that it is available in any emergency. If it is necessary to use the room as a workplace, procedures should be in place to ensure that its use as a first-aid room is not prejudiced.

What records in respect of first-aid treatment, etc. need to be kept by the employer?

It is considered good practice to keep records of all incidents that require any first-aid treatment or attendance by a first aider.

A record book should be available to record the following:

- the date, time and place of the incident
- the injured person's name and job title
- a description of the person's injuries or illness
- details of first-aid treatment given
- details of any actions taken after treatment given (e.g. taken to hospital or went home)
- the name and signature of the person who gave the first-aid treatment or who oversaw the incident.

Consideration must be given to any data protection requirements and for the anonymity of personal information. It would be sensible to use a new page for each person treated and for previous entries to be kept in a secure drawer. People should not be able to flip through the first-aid record book and learn personal facts about colleagues or others.

Case study

A young employee arrived at work in a call centre and after about an hour reported that she did not feel well. She had a headache and felt sick. The supervisor called the first aider, who suggested that the employee would probably benefit from a lay down in the first-aid room. She was taken to the first-aid room and settled down on the couch, still feeling quite poorly.

Her colleagues were quite busy and forgot all about the young woman being in the first-aid room until about lunchtime, when a colleague went to visit her and found that she was in a coma. An ambulance was called and the young woman was taken to hospital. Unfortunately, she died later that day from a brain haemorrhage.

Could anything have been done differently?

Yes. The first aider who attended the young woman should have been responsible for her care while she was in the first-aid room and should have visited her every 30 minutes or so. If the woman had not felt better after, say, an hour, she should have been taken to her GP or to hospital. Records should have been kept of the checks carried out on her. The woman might still have died but she should not have been left unattended for several hours.

FIRST AID – RISK ASSESSMENT

The aim of the first aid risk assessment process is to assist in the determination of the appropriate first-aid facilities and number of trained first aiders for a specific area (i.e. department, building and/or site).

The responsible person must consult with the staff of the area on all aspects of the provision of first aid (this is suggested to be through departmental/area health and safety committee meetings).

Department: . . . .

Location: . . . .

Director: . . . .

Departmental safety advisor: . . . .

Date: . . . .

Assessors: . . . .

. . . .

Current number of first aiders and level of training: . . . .

Approximate number of people working between 8.30 a.m. and 6.30 p.m.: . . . .

Approximate number of staff in the area outside work hours (including weekends): . . . . . .

Is lone working carried out in the department? . . . .

Is the department spread out (e.g. are there several buildings on the site or multi-floor buildings)? . . . .

Distance to nearest major hospital: . . . .   Travel time: . . . .

Maximum distance to location of existing first-aid kits: . . . .

Summary of major incidents over the last 12 months: . . . .

. . . .

. . . .

Do you have any work-experience trainees, volunteers or honorary staff? . . . .

. . . .

Hazards

List specific hazards in the area (e.g. slips and trips, working at height, plant or machinery, moving objects, electricity, radiation, chemicals, dust, manual handling).

Hazard	Location

Are there hazards or health concerns for which an extra first-aid kit or specialised treatment is required (e.g. chemicals, potential for burn, eye injuries, field trips)?

Hazard	Location

Recommendations

Contents of kit	
No. and location of kits	
No. of first aiders required	
No. of appointed persons required	
Additional recommendations	

Action list

Item for action	Completion date	Date signed off	Initials
Signed:			
Printed name:			
Date:			

Risk Assessments: Questions and Answers
ISBN 978-0-7277-6076-0

ICE Publishing: All rights reserved
http://dx.doi.org/10.1680/raqa.60760.227

Chapter 14
Work-related limb disorders

What are the steps for an effective risk assessment for work-related limb disorders?

Risk assessment for work-related limb disorders is best carried out in two stages:

- identify problem tasks
- assess the risks.

It is essential to identify all problem, or likely problem, tasks in the organisation.

Find out which tasks people do. Walk the shop floor, office, department store, etc., and watch what people do.

Are people carrying out repetitive tasks quickly, such as at supermarket checkouts, or are the tasks less frequent but involve heavier weights or more awkward postures?

Record the basic features of the task on a checklist.

Review sickness and absence records.

Talk to employees – do they have any pain in their limbs, have general aches, etc.?

How long do people carry out tasks without stopping?

What is a work-related upper limb disorder?

Work-related upper limb disorders (WRULDs) are conditions that affect the:

- neck
- shoulders
- arms
- elbows
- wrists
- hands
- fingers.

227

The symptoms include:

- aches and pains
- difficulty in moving
- swelling
- stiffness of joints.

Upper limb disorders can be recognised industrial injuries, and include:

- carpel tunnel syndrome
- tendonitis
- bursitis
- trigger finger
- vibration white finger.

Are work-related upper limb disorders the same as repetitive strain injuries?

Yes. Generally anything that causes symptoms to the upper limbs and is caused by repeatedly undertaking the same tasks in the same way will fall into the category of upper limb disorders or repetitive strain injuries.

A wider term for such injuries in 'musculoskeletal disorders'.

What parts of the body are affected?

Usually the neck, shoulders, arms, wrists, hands and fingers are affected. In particular, the muscles, tendons, ligaments, nerves or soft tissue associated with these joints.

What causes work-related upper limb disorders?

Upper limb disorders can occur in jobs that require repetitive finger, hand or arm movements, twisting movements, squeezing, hammering or pounding, pushing, pulling, lifting or reaching movements.

Both office-based jobs and manual activities can lead to WRULDS. For example:

- repetitive assembly line work
- inspection and packing
- meat and poultry preparation
- keyboard work
- supermarket checkouts
- construction work.

The risk of WRULDs is higher when there is:

- repetitive work processes
- force
- awkward posture
- insufficient rest breaks to allow muscles, etc. to recover.

As an employer, do I need to complete a risk assessment to establish whether employees are exposed to work-related upper limb disorders?

Yes. A health and safety risk assessment needs to be done for any job task or activity where there is a risk to an employee (or others) of a WRULD.

Upper limb disorders are now quite well known, and employers should be aware that they affect employees who carry out repetitive jobs of all types.

The general risk assessment undertaken for all work activities as required by the Management of Health and Safety at Work Regulations 1999 will suffice for identifying the risk of WRULDs occurring.

There is no specific legislation that deals with WRULDs, but the Health and Safety (Display Screen Equipment) Regulations 1992 require specific risk assessments to be carried out for work with display screens and associated equipment.

How should work-related upper limb disorders be tackled?

The Health and Safety Executive has published a guidance booklet on upper limb disorders (*Upper Limb Disorders in the Workplace*, HSG60, 2002) and advocates a seven-stage management approach as follows:

Step 1: Understand the issues and commit to action
Step 2: Create the right organisational environment
Step 3: Assess the risks of WRULDs in the workplace
Step 4: Reduce the risks of WRULDs
Step 5: Educate and inform the workforce
Step 6: Manage any incidents of WRULDs
Step 7: Carry out regular audits and monitor the programme's effectiveness.

Step 1: Understand the issues and commit to action

Everybody should understand that job tasks or activities could cause upper limb disorders, and know how to identify the symptoms, report occurrences, adapt working processes, etc.

Managers need to recognise that WRULDs are a significant hazard in the workplace and that they have a responsibility to reduce or eliminate them. Employees off sick with WRULDs can be expensive, as can temporary replacements, inefficient working due to lower performance, etc.

Dealing with WRULDs is not necessarily expensive.

Operate a 'zero tolerance' policy on WRULDs.

Step 2: Create the right organisational environment
Create an environment where managers are seen to recognise the hazards, risks and costs of poor working practices that lead to WRULDs.

Involve employees in seeking their own solutions.

Produce policies and procedures on how to undertake jobs, when to have breaks, acceptable behaviour, etc.

Ensure that employees who complain they are suffering symptoms are not ridiculed. Listen to concerns and create an environment that shows you are prepared to act and resolve issues.

Step 3: Assess the risks of WRULDs in the workplace
Carry out regular and effective risk assessments on job tasks. Make sure that the persons carrying out the risk assessments are competent.

Step 4: Reduce the risks of WRULDs
A process of risk reduction should be undertaken using a number of approaches (e.g. ergonomics).

Endeavour to eliminate the risk of injuries through change working practices, tools, posture, body positions, etc.

Involve work groups in seeking solutions – often worker participation brings 'buy in' to solutions and they become more effective.

Review accident and sickness records. Are the same or more people off sick with limb disorders, pain, etc. Do people have just 1 or 2 days off and use 'pain in my wrist' or similar as the reason for absence.

Step 5: Educate and inform the workforce
Provide information, instruction and training to employees.

What are the hazards and repetitive tasks? What are the symptoms? How can employees improve job processes, etc.?

Ensure employees know about the risk assessments and the proposed physical control measures.

Step 6: Manage any incidents of WRULDs

Respond to any complaints of limb injuries or disorders. Review the work in hand for the individual. Change things immediately.

WRULDs need not be permanent and can be reduced or eliminated if early action is taken. Continued strain on already strained muscles and tendons, for example, will make the injury worse, whereas stopping the task as soon as the pain is identified will enable the muscles or tendons to repair themselves and heal.

Step 7: Carry out regular audits and monitor the programme's effectiveness

As with any health and safety management programme, the control measures introduced need to be audited and reviewed to ensure their effectiveness.

Regular monitoring should take place to check that employees are following the correct procedures, etc.

Any deviations, new tools, adapted tools, etc. need to be assessed to establish why they are needed and how they can be used safely, etc.

What are lower limb disorders?

Lower limb disorders affect the legs and feet, from hips to toes. About 80% of the damage to the hips, knees and legs at work is due to overuse. Workers may report lower limb pain, aching and numbness without a specific disease being identified or present.

What causes lower limb disorders?

Evidence suggests that there are several recognised diseases of the lower limb that can be work related, such as hip and knee osteoarthritis, knee bursitis, meniscal lesions and tears, stress fractures and reaction injuries, and varicose veins of the lower legs.

What are the symptoms of lower limb disorders?

The symptoms are pain, tenderness or stiffness of a *joint*, and an inability to straighten or bend the joints. Other symptoms are tenderness, aches and pains, stiffness, weakness, tingling, numbness, cramp and swelling to the *muscles* of the lower limbs.

These symptoms may suggest overuse or some underlying condition, and, in the first instance, employees should seek advice from their GP.

How can I prevent lower limb disorders?

Because most injury happens as a result of overuse, the most effective way to reduce the risk is to avoid overuse by providing mechanical aids or rotating duties to reduce the time spent carrying out a 'risky' task and give time for recovery. Where possible, provide seating rather than requiring squatting or kneeling, and ensure that employees take regular breaks. Provide personal protective equipment such as:

- knee pads to protect the knee while kneeling on hard floor surfaces, to improve comfort and prevent lacerations and penetrating injuries
- anti-fatigue matting may be effective in reducing the risks from prolonged standing, but using the mats in the workplace requires careful consideration because of the increased risk of slips and trips
- shock-absorbing insoles and modified shoes have positive effects mainly in respect of injuries from vertical impact loads, but there is limited evidence that using such insoles reduces the risk of developing a lower limb disorder.

How can working in a standing position affect a person's health?

A person's body is affected by the arrangement of the work area and by the tasks that he or she does while standing. The layout of the workstation, the tools used and the placement of keys, controls and displays that the worker needs to operate or observe will determine, and as rule limit, the body positions that the worker can assume while standing. As a result, the worker has fewer body positions to choose from, and the positions themselves are more rigid. This gives the worker less freedom to move around and to rest working muscles. This lack of flexibility in choosing body positions contributes to health problems.

These conditions commonly occur where the job is designed without considering the characteristics of the human body. When job design ignores the basic needs of the human body (and individual workers), work can cause discomfort in the short term and can eventually lead to severe and chronic health problems.

What are some of the principles of good workplace design to reduce lower limb disorders?

In a well-designed workplace the worker has the opportunity to choose from a variety of well-balanced working positions and to change between them frequently.

Work tables and benches should be adjustable. Being able to adjust the working height is particularly important to match the workstation to the worker's individual body size and to the worker's particular task. Adjustability ensures that the worker has an opportunity to carry out work in well-balanced body positions. If the workstation cannot be adjusted, platforms to raise the shorter worker or pedestals on top of workstations for the tall worker should be considered.

Organisation of the work space is another important aspect. There should be enough room to move around and to change body position. Providing built-in foot rails or portable footrests allows the worker to shift their body weight from one leg to the other. Elbow supports for precision work help reduce tension in the upper arms and neck. Controls and tools should be positioned so the worker can reach them easily and without twisting or bending.

Where possible, a seat should be provided so that the worker can do the job either sitting or standing. The seat must place the worker at a height that suits the type of work being done. For work that requires standing only, a seat should be provided in any case, to allow the worker to sit occasionally. Seats at the workplace expand the variety of possible body positions and give the worker more flexibility.

The benefits from greater flexibility and a variety of body positions are twofold. The number of muscles involved in the work is increased, which equalises the distribution of loads on different parts of the body. There is, therefore, less strain on the individual muscles and joints used to maintain the upright position. Changing body positions also improves blood supply to the working muscles. Both effects contribute to the reduction of overall fatigue.

Quality of footwear and type of flooring material are also major factors contributing to standing comfort.

All of the above should be considered in the risk assessment and the most suitable controls implemented.

Remember that one size will not fit all employees, and the most successful approach is to foster worker engagement and flexibility in adopting solutions.

What are some of the control measures that may help reduce work-related limb disorders?

Control measures to reduce or eliminate WRULDs can be extremely wide ranging and are best judged separately for each job task or activity.

Any improvement in reducing repetitive jobs, awkward postures, carrying of weights, etc. will help reduce risks of injury.

Some control measures that might be considered are:

- mechanise or automate the process
- modify the operation or process
- change the shape of tools
- change the height or the style of chairs
- reduce packaging size
- reduce weights
- reduce reach distances
- lower shelving
- provide mechanical aids
- reduce 'double handling'
- move tasks nearer to stock
- improve the environment
- create more work space
- give employees some control over their jobs
- introduce multi-tasking
- rotate employees
- adapt the job to the worker, not the worker to the job
- provide soft-grip tools
- review personal protective equipment
- use lightweight tools
- improve lighting and other working environment factors.

Identified upper limb disorders, occupational risk factors and symptoms

Disorder	Occupational risk factors	Symptoms
Tendonitis/ tenosynovitis	Repetitive wrist motions Repetitive shoulder motions Sustained hyperextension of arms Prolonged load on shoulders	Pain, weakness, swelling, burning sensation or dull ache over the affected area
Epicondylitis (elbow tendonitis)	Repeated or forceful rotation of the forearm and bending of the wrist at the same time	As above
Carpal tunnel syndrome	Repetitive wrist motions	Pain, numbness, tingling, burning sensations, wasting of muscles at base of thumb, dry palm
De Quervain's disease	Repetitive hand twisting and forceful gripping	Pain at base of thumb
Thoracic outlet syndrome	Prolonged shoulder flexion Extending arms above shoulder height Carrying loads on the shoulder	Pain, numbness, swelling of the hands
Tension neck syndrome	Prolonged restricted posture	Pain

WORK-RELATED UPPER LIMB DISORDERS – INITIAL ASSESSMENT

Company: ....

Department: ....

Initial assessment completed by: ....

Date: ....

Job task being assessed: ....

Describe work activity/process: ....

....

....

....

....

Does the job involve:

	Yes	No
Gripping tools?		
Squeezing tools, implements?		
Twisting?		
Reaching?		
Moving things?		
Lifting things?		
Finger/hand movement?		
Fast, short movements?		
Awkward posture?		
The use of force?		
Repetition of task?		

Does the employee:

	Yes	No
Have control of their work flow?		
Have to work to the speed of others?		
Take regular breaks?		

Are there:

	Yes	No
Actual cases of WRULDs?		
Complaints from employees about pains in the hands, wrists, arms, etc.?		
Home-made improvised changes to equipment, workstations or tools?		

Any other comments:

. . . .

. . . .

. . . .

Is a FULL RISK ASSESSMENT REQUIRED (this will be YES if any of the above questions/statements were answered 'yes'):   Yes/No

WORK-RELATED UPPER LIMB DISORDERS – FULL RISK ASSESSMENT

Name of company: . . . .

Name of assessor: . . . .

Date of assessment: . . . .

Name of employee being assessed: . . . .

Job title: . . . .

Job task or activity: . . . .

Describe task in detail: . . . .

. . . .

. . . .

. . . .

Repetition
Does task involve repetition?   Yes/No

Describe, including frequency of movement, etc.

. . . .

. . . .

. . . .

Is repetition continuous for the shift or interspersed with breaks?

. . . .

Is the job task always the same?   Yes/No

What could be done to reduce repetition?

. . . .

. . . .

. . . .

Working posture

	Yes	No
Is the wrist bent in any way?		
Does the posture look awkward?		
Is a grip needed?		
Does the hand easily span the distance?		
Are tools awkward to hold, handle?		
Are tools turned, twisted?		
Are tools or equipment too high for the worker?		
Are tools or equipment too low for the worker?		
Can the worker reach things?		
Is there a lot of overhead reaching?		
Are weights held with outstretched arms?		
Do the head and neck have to be put in awkward positions?		
Is there any visual sign of discomfort?		

What could be done to improve the working posture?

. . . .

. . . .

Force
Does the work activity require force of any kind, including pushing, pinching, twisting, hammering, need to hold item against the body, etc.?

Describe any type of force noted.

. . . .

. . . .

. . . .

What can be done to reduce the need for force?

. . . .

. . . .

. . . .

Working environment

Describe the working environment, heating, lighting, ventilation, overcrowding, obstructions, noise, confined spaces, cramped working space, etc.

Does anything add to the discomfort of the employee or create the need for more repetition of job tasks?

. . . .

. . . .

. . . .

What can be done to improve the work environment?

. . . .

. . . .

. . . .

Psycho-social factors

What factors may affect the wellbeing of employees, and what could contribute to poor working practices, repetition of tasks, etc.?

. . . .

. . . .

. . . .

Do employees have control of their work pattern?	Yes/No
Do they have to work to the speed of others?	Yes/No
Can they have breaks?	Yes/No
Are they on piece work?	Yes/No

What can be done to improve conditions?

. . . .

. . . .

. . . .

What other factors could affect the health, safety and wellbeing of the employee or the person carrying out this task?

. . . .

. . . .

. . . .

ACTION PLAN

What needs to be done?	By whom?	When?
Risk assessment reviews		
Completion of action plan		
General review		

Signed: . . . .

Position: . . . .

Date: . . . .

Risk Assessments: Questions and Answers
ISBN 978-0-7277-6076-0

http://dx.doi.org/10.1680/raqa.60760.243

Chapter 15
Design risk assessments

What is a design risk assessment?

A design risk assessment is usually a record of the hazards and risks that have been identified during the process of designing a structure or building, and the control measures that the designer deems appropriate to mitigate the risks to acceptable levels.

The Construction (Design and Management) Regulations 2015 (CDM 2015) refer to the need to consider the 'principles of prevention' during the design process. Designers have to be able to demonstrate how they have achieved this and keeping a record of this is recommended. The record is referred to as the 'design risk assessment', and it can be of various styles and content.

Designers have duties under CDM 2015 to avoid foreseeable risks to the health and safety of people constructing their design, affected by the construction works or using or resorting to the building or structure once built. What does this mean?

Designers must consider the health and safety aspects of their designs and must avoid foreseeable risks – those risks that their experience, knowledge and competence tells them they should know about.

Many accidents occur in buildings because the end design is not user friendly, or the materials looked good when the structure was first built but after some wear and tear have become hazardous, or the access to plant and equipment is poor and maintenance personnel have to undertake hazardous tasks to do their jobs, etc.

Designers have a great influence on the future safe use of the building, and the CDM 2015 require them to recognise this fact by considering health and safety.

Designers can also have a great influence on the safety of contractors by paying attention to health and safety when they choose materials, building location, site set-out, access, construction sequences, etc.

Designers can do more to ensure that contractors and their operatives work more safely on construction sites by ensuring that they specify more appropriate materials, equipment, etc. (e.g. pay greater attention to size and weight).

If the risk is foreseeable (i.e. you could see or know about it beforehand), then the CDM 2015 require the designer to avoid it.

Risks may be foreseeable but they may not always be avoidable. What happens then?

Risks will always be with us and it will be impossible to eliminate all risk. Health and safety legislation is qualified by the term 'as far as is reasonably practicable', and this means that a judgement can be made.

If a risk cannot be eliminated completely, it may be possible to replace it with a lower risk by using something that has a lower associated risk. For example, a substance known to cause cancer could be replaced with one known to cause skin irritation. There would still be a risk but the consequences of using the substance will be less and perhaps more acceptable.

Health and safety law refers to:

- the hierarchy of risk control
- principles of prevention.

Designers would have to demonstrate that whenever they are unable to eliminate a hazard and the associated risk they have ensured that they have chosen a design with the minimum risk or that they have protected against the hazard and risk.

What is safe design?

Safe design is the integration of hazard identification and risk assessment methods to eliminate or minimise the risks of injury throughout the life of a product or structure.

The safe-design approach begins with an emphasis on making choices about design, materials and methods of manufacture or construction to enhance the safety of the finished product.

Safe design is good business because it improves safety and reduces costs by:

- improving risk management of workplace health and safety issues during the construction phase
- protecting constructors from injury
- reducing the need for redesign and retrofitting.

It should begin at the design phase of a project and not be left until construction starts.

Designers, therefore, have the greatest opportunity to influence the safety of the structure and the process of construction.

Is there a recognised format for design risk assessments?

No. The CDM 2015 do not specify an exact format that has to be used for design risk assessments. They require merely that hazards and risks in the design are considered and, where appropriate, recorded, so that the information can be passed to other members of the design team, the principal designer, the principal contractor and all other contractors, as appropriate.

Suitable information on hazards and risks can be annotated by drawings if this will give clear guidance to whoever will be reading the drawings. Information can also be collated in pro-forma design risk assessment forms – although the Health and Safety Executive (HSE) states that complicated design risk assessments are not expected.

If drawings are designed using computer-aided design (CAD) systems, text can be annotated by drawings to convey the hazards and risks and proposed control measures.

Information on hazards and risks associated with the design of the structure, its construction, and subsequent use and maintenance must be freely available to all members of the project team.

No one has a monopoly on the safest way to do things, and the previous experience of other team members may add a valuable contribution to the debate on health and safety.

Some clients and principal designers advocate that a project risk register is collated – usually by the principal designer – and that all hazards, risks and solutions are recorded, with dates of actions taken, changes made, etc. Common and expected hazards are not expected to be laboriously recorded – competent contractors should be aware of the dangers of using ladders to work at height. What is more important in the design risk assessment is the thought process the designer has undertaken to see if works can be undertaken at ground level, thus negating the need for ladders and eliminating a site risk. Could staircases be installed early on, thereby reducing the use of ladders as a means of access to upper levels? The designer can influence this process, and it forms part of the design risk assessment.

What is 'initial design work'?

Initial design work is generally taken to include:

- an appraisal of the project needs
- the setting of project objectives
- the feasibility of objectives in relation to costs
- possible constraints on the project
- possible desktop studies for contaminated land and remedial works
- likely procurement methods
- the strategic brief
- confirmation of key project team members (positions rather than names).

The Royal Institute of British Architects suggest that stages A and B in their Plan of Work are initial design work.

Once drawings and specifications have been drawn up and planning permission is being sought the design has moved into 'outline design', and the principal designer should definitely have been appointed.

As designers, must we provide information on all hazards and risks associated with the project?

No. The CDM 2015 require designers to take account of significant risks that competent contractors might *not* be aware of in respect of their designs.

All competent contractors, for instance, should be aware of the hazards and risks of working at heights, and the designer does not need to give the principal contractor chapter and verse on the safety precautions for working at heights. However, the designer should consider how to *reduce* the need to work at height, and if a particular sequence of construction is envisaged the designer must provide this information to the principal contractor, via the principal designer on notifiable projects and directly to the contractor on other projects.

Providing too much information on all hazards and risks of construction obscures the important detail about significant risks.

Contractors need to know about any specific materials specified or construction sequences planned that they may not be overly familiar with (e.g. installation of glazed atria) in order to achieve the desired design effect in the building.

Do the Construction (Design and Management) Regulations 2015 specifically require design risk assessments to be generated on every construction project?

No. The CDM 2015 do not specifically state that design risk assessments have to be produced. The regulations state that designers must manage design risks by avoiding foreseeable risks and by reducing residual risks so far as is reasonably practicable by following the principles of prevention.

To comply with their duties under the CDM 2015 designers must take reasonable steps to provide with their designs sufficient information about aspects of the design of the structure or its construction or maintenance (including cleaning) as will adequately assist:

- clients
- other designers
- contractors.

The information that designers have to provide to other duty holders can be in any format but it should be brief, clear, precise and project specific.

Information can be in the form of notes on drawings, written information provided with the design drawings, formal risk assessment forms or project specific instructions.

HSE inspectors are not looking for meaningless paperwork – they want to see evidence that designers have thought about health and safety issues and consequences during the development of the project and have done everything that they can to assist in reducing on-site accidents and ill health, and have thereby contributed to a reduction in accidents involving maintenance and cleaning workers and the building users.

The principles of prevention are:

- avoiding risks
- evaluating the unavoidable risks
- combating the risks at source
- adapting the work to the individual
- adapting to technical progress
- replacing the dangerous with the non/less dangerous
- developing a prevention policy
- collective and individual protection
- giving appropriate instruction.

What are some of the things I need to think about regarding safety in design?

All designers should have procedures in place to help them review and record the processes they follow to ensure that their designs follow the principles of prevention in respect of health and safety. The following ideas can be adopted by designers on a wide range of projects.

Design for safe construction

Control measures for risks relating to the construction of a structure include:

- Providing adequate clearance between the structure and overhead electrical lines by burying, disconnecting or rerouting cables before construction begins, to avoid 'contact' when operating cranes and other tall equipment.
- Designing components that can be prefabricated off site or on the ground to avoid assembling or erecting at height and to reduce worker exposure to falls from heights or being struck by falling objects (e.g. fixing windows in place at ground level prior to the erection of panels).
- Designing parapets to a height that complies with guardrail requirements, eliminating the need to construct guardrails during construction and future roof maintenance.
- Using continual support beams for beam-to-column double connections, by adding a beam seat, an extra bolt hole or other redundant connection points during the connection process. This will provide continual support for beams during erection, to eliminate falls due to unexpected vibrations, misalignment and unexpected construction loads.
- Designing and constructing permanent stairways to help prevent falls and other hazards associated with temporary stairs and scaffolding, and scheduling these at the beginning of construction.
- Reducing the space between roof trusses and battens to reduce the risk of internal falls during roof construction.
- Choosing construction materials that are safe to handle.
- Limiting the size of prefabricated wall panels where site access is restricted.
- Selecting paints or other finishes that have low volatile organic compound emissions.
- Indicating, where practicable, the position and height of all electric lines, to assist with site safety procedures.

Design to facilitate safe use

Consider the intended function of the structure, including the likely systems of use, and the type of machinery and equipment that may be used.

- Consider whether the structure may be exposed to specific hazards, such as manual tasks in health facilities, occupational violence in banks or dangerous goods storage in warehouses.
- Risks relating to the function of a structure can be controlled by designing traffic areas such that they separate vehicles and pedestrians.
- Using non-slip materials on floor surfaces in areas exposed to the weather or in dedicated wet areas.

- Providing sufficient space to safely install, operate and maintain plant and machinery.
- Providing adequate lighting for intended tasks in the structure.
- Designing spaces that accommodate or incorporate mechanical devices, to reduce manual task risks.
- Designing adequate access (e.g. allowing wide enough corridors in hospitals and nursing homes for the movement of wheelchairs and beds).
- Designing effective noise barriers and acoustic treatments to walls and ceilings.
- Specifying plant with low noise emissions or designing the structure to isolate noisy plant.
- Designing floor loadings to accommodate heavy machinery that may be used in the building, and clearly indicating on documents the design loads for the different parts of the structure.

Design for safe maintenance

Risks relating to cleaning, servicing and maintaining a structure can be controlled by:

- designing the structure so that maintenance can be performed at ground level or safely from the structure, for example, positioning air-conditioning units and lift plant at ground level, designing inward opening windows, integrating window cleaning bays or gangways into the structural frame
- designing features to avoid dirt traps
- designing and positioning permanent anchorage and hoisting points into structures where maintenance needs to be undertaken at height
- designing safe access, such as fixed ladders, and sufficient space to undertake structure maintenance activities
- eliminating or minimising the need for entry into confined spaces
- using durable materials that do not need to be re-coated or treated.

Modification of a structure or its use

Design is not always focused on the generation of an entirely new structure. It can involve the alteration of an existing structure, which may require partial or complete demolition.

Any modification of a structure requires reapplication of the processes detailed in the design phases. Consultation with professional engineers or other experts may be necessary in order to assess the impact of any proposed modifications or changes in design (e.g. changes in the load spread across a building floor when heavy equipment is relocated, modified or replaced).

This ensures that any new hazards and risks are identified and controlled, and that the safety features already incorporated in the design are not affected. Additional design issues identified in these phases should be passed back to the designer.

Demolition and dismantling

A structure should be designed to enable demolition using existing techniques. The designer should provide information so that potential demolishers can understand the structure, load paths and any features incorporated to assist demolition, as well as any features that require unusual demolition techniques or sequencing.

Designers of new structures are well placed to influence the ultimate demolition of a structure by designing-in facilities such as lifting lugs on beams or columns and protecting inserts in pre-cast panels so that they may be used for disassembly. Materials and finishes specified for the original structure may require special attention at the time of demolition, and the client should be advised of any special requirements for the disposal and/or recycling of such materials or finishes through the risk assessment documentation.

DESIGN RISK ASSESSMENT – HSE INSPECTOR'S CHECKLIST

Design stage	What to look for	Basis for intervention (i.e. what action to take)
Prior to concept selection (at or before project sanction)	Policy for safety in design Criteria for concept selection (e.g. life-cycle aspects considered) Roles, responsibilities and competence of relevant personnel Adequacy of health and safety advice	Sector specific – general advice
Detailed design	As above, and: Application of relevant and current good practice Effective approval processes, including interaction with risk assessment Application of formal and structured risk assessments and their effectiveness in reducing risks Effective change control procedures Development of appropriate information to enable safe operation, maintenance and repair	Sector-specific legislation Codes, standards and other good practice
Construction	As above, and: Effective material control Effective quality control (e.g. leak testing, welding procedures) Conformity to design	Sector-specific and general legislation (e.g. CDM 2015) Codes, standards and other good practice

DESIGN RISK ASSESSMENT – DESIGNER'S CHECKLIST

The following list may be used to assist in identifying hazards and controlling risks associated with the design of a structure throughout its life cycle.

ELECTRICAL SAFETY
- Earthing of electrical installations
- Location of underground and overhead power cables
- Protection of leads/cables
- Number and location of power points

FIRE AND EMERGENCIES
- Fire risks
- Fire detection and fire-fighting
- Emergency routes and exits
- Access for and structural capacity to carry fire tenders
- Other emergency facilities

MOVEMENT OF PEOPLE AND MATERIALS
- Safe access and egress, including for people with disability
- Traffic management
- Loading bays and ramps
- Safe crossings
- Exclusion zones
- Site security

WORKING ENVIRONMENT
- Ventilation for thermal comfort and general air quality, and specific ventilation requirements for the work to be performed on the premises
- Temperature
- Lighting, including of plant rooms
- Acoustic properties and noise control (e.g. noise isolation, insulation and absorption)
- Seating
- Floor surfaces to prevent slips and trips
- Space for occupants

PLANT
- Tower crane locations, loading and unloading
- Mobile crane loads on slabs
- Plant and machinery installed in a building or structure

- Materials handling plant and equipment
- Maintenance access to plant and equipment
- The guarding of plant and machinery
- Lift installations

AMENITIES AND FACILITIES
- Access to various amenities and facilities, such as storage, first-aid rooms/sick rooms, rest rooms, meal and accommodation areas, and drinking water

EARTHWORKS
- Excavations (e.g. risks from earth collapsing or engulfment)
- Location of underground services

STRUCTURAL SAFETY
- Erection of steelwork or concrete frameworks
- Load-bearing requirements
- Stability and integrity of the structure

MANUAL TASKS
- Methods of material handling
- Accessibility of material handling
- Loading docks and storage facilities
- Workplace space and layout to prevent musculoskeletal disorders, including facilitating use of mechanical aids
- Assembly and disassembly of prefabricated fixtures and fittings

SUBSTANCES
- Exposure to hazardous substances and materials, including insulation and decorative materials
- Exposure to volatile organic compounds and off-gassing through the use of composite wood products or paints
- Exposure to irritant dust and fumes
- Storage and use of hazardous chemicals, including cleaning products

FALLS PREVENTION
- Guard rails
- Window heights and cleaning
- Anchorage points for building maintenance and cleaning
- Access to working spaces for construction, cleaning, maintenance and repairs
- Scaffolding
- Temporary work platforms

■ Roofing materials and surface characteristics, such as fragility, slip resistance and pitch

SPECIFIC RISKS
■ Exposure to radiation (e.g. electromagnetic radiation)
■ Exposure to biological hazards
■ Fatigue
■ Working alone
■ Use of explosives
■ Confined spaces
■ Over- and under-water work, including diving and work in caissons with compressed-air supply

NOISE EXPOSURE
■ Exposure to noise from plant or from the surrounding area

Risk Assessments: Questions and Answers
ISBN 978-0-7277-6076-0

ICE Publishing: All rights reserved
http://dx.doi.org/10.1680/raqa.60760.255

Chapter 16
Water risk assessments for *Legionella*, Legionnaires' disease and other hazards

What is Legionnaires' disease?

Legionnaires' disease is a respiratory disease caused by the inhalation of the *Legionella* bacterium, *Legionella pneumophila*. There are other strains of bacteria that also cause the disease or an illness similar to it (i.e. Pontiac fever).

The symptoms are:

- flu-like
- malaise
- general pains
- headache
- raised temperature – often up to 40°C
- dry cough
- nausea, vomiting, diarrhoea (at times)
- severe respiratory infection.

The onset of the symptoms varies, and the common incubation period is 2–10 days.

Is everybody affected to the same extent?

No, not necessarily. Certain groups in the population are more susceptible to the disease than others. Healthy people will usually recover from the illness once antibiotic treatment has been given.

Those most at risk are people with conditions that suppress the immune system (e.g. cancer, respiratory diseases and kidney disease) or diabetes, and people who smoke or have a chronic dependent illness (e.g. alcoholism).

The age range of those most commonly affected is 40–70 years.

Where are *Legionella* bacteria found?

Legionella bacteria are widespread in natural watercourses, including rivers, streams and ponds. The bacterium can also be found in soil, although it is far more prevalent in water systems.

Commonly, *Legionella* bacteria are found in hot and cold water systems, particularly those that store water and recirculate it (e.g. air-conditioning systems and spa baths).

How is the infection spread?

The *Legionella* bacterium is spread by airborne droplets of water or water vapour (e.g. sprays and mists).

The water droplets in spray or mist form are inhaled by individuals and the infection takes hold within the body. The number of bacteria that need to be inhaled in order to develop the symptoms is not defined – it will vary from person to person depending on their general state of health.

Are outbreaks of Legionnaires' disease common?

No, not especially, although when they do occur they can affect large amounts of people.

For an outbreak of Legionnaires' disease to occur a sequence of events has to take place:

- conditions have to exist that suit the multiplication of the bacteria
- the water temperature has to be between 20°C and 45°C
- sludge, scale, rust, algae or other organic matter must be present to provide the bacteria with nutrients
- there needs to be a means of creating breathable droplets of water
- there must be contact with the infected droplets by a susceptible person.

Cooling towers are likely to be the source of many outbreaks, and there are specific requirements on dutyholders to ensure that risks are controlled.

Case study

Legionnaires' disease outbreak in Scotland, 2012

In 2012, an outbreak of Legionnaires' disease affected a defined population in south-west Edinburgh. It had considerable impact on NHS services during June 2012. Over 1000 patients were investigated and treated in primary care. Forty-five of the confirmed cases were admitted to acute hospitals in NHS Lothian. Twenty-two patients required admission to critical care; 19 were admitted to intensive care and three to a high dependency unit. In total, 92 cases were identified in the outbreak; 56 confirmed cases and 36 probable or possible cases. Seven of the confirmed cases and two in the probable or possible category were identified outwith NHS Lothian.

Four deaths were reported among formally confirmed cases. The case fatality rate was 7.1% of confirmed cases and 4.3% of all cases.

The cause of the outbreak was likely to have been cooling towers serving businesses in the area, although no actual *Legionella* bacteria were isolated from samples taken by the investigating authorities.

What are an employer's legal responsibilities?

The general duties of the Health and Safety at Work etc. Act 1974 apply; in particular, Section 2 – the duty to ensure a safe place of work.

In addition, the Management of Health and Safety at Work Regulations 1999 and the Control of Substances Hazardous to Health Regulations 2002 require employers to consider the hazards and risks to employees from work activities, including any exposure to biological organisms.

Employers must complete risk assessments for the likely exposure of employees and others to *Legionella* bacteria. If there is a risk of exposure, the employer must implement control measures. Approved codes of practice have been issued on the prevention and control of legionellosis (including Legionnaires' disease) as well as a guidance on the control of legionellosis including Legionnaires' disease (Legionnaires' Disease. The Control of Legionella Bacteria in Water Systems. Approved Code of Practice and Guidance (L8, 4th edn, 2013), Legionnaires' Disease. Technical Guidance (HSG274, 2014), Legionnaires' Disease. A Brief Guide for Dutyholders (NDG458, 2012)).

Employers also have duties under the Notification of Cooling Towers and Evaporative Condensers Regulations 1992. These regulations require that local authorities are notified of the location of wet cooling towers and evaporative condensers so that, if there is an outbreak, the control infection team or environmental health department knows where to start its investigations for possible sources of contamination.

Also, if considered appropriate, the local authority can require employers or others to control or abate a 'statutory nuisance' if it believes a breach of the Environmental Protection Act 1990 has occurred. Any 'fumes or gases emitted from premises' and 'dust, steam, smell or other effluvia arising on industrial, trade or business premises' that is considered prejudicial to health or a nuisance can be the subject of enforcement action.

Employers, therefore, need to be mindful of environmental legislation as well as health and safety.

Who is a 'responsible person' and who can be appointed as the responsible person?

The responsible person will take day-to-day responsibility for managing the control of any identified risk from *Legionella* bacteria. Employers and others are advised to ensure that they allocate specific responsibilities for the management of *Legionella* risks to a named individual or, where there are significant risks, more than one individual.

Anyone can be appointed as the responsible person as long as they have sufficient authority, competence, skills and knowledge about the building and its services installations to ensure that all operational procedures are carried out in a timely and effective manner and to implement the control measures and strategies (i.e. they are suitably informed, instructed, trained and assessed). They should be able to ensure that tasks are carried out in a safe and technically competent manner.

If a dutyholder is self-employed or a member of a partnership, and is competent, they may appoint themselves. The responsible person should be suitably informed, instructed and trained and their suitability assessed. They should also have a clear understanding of their duties and the overall health and safety management structure, and policy in the organisation.

What do employers need to do to reduce the risk of a *Legionella* outbreak?

The first step is to carry out a risk assessment to identify whether the hazard is present on the premises. All water supplies have the potential to be contaminated with *Legionella* bacteria, so, mostly, the hazard cannot be eliminated at source.

Consider whether your employees and others are exposed to:

- hot water
- warm water (about 20°C)
- showers
- air-conditioning cooled by water
- humidifiers
- leisure facilities (e.g. spa baths and Jacuzzis)
- hot tubs
- water-spraying operations.

Members of the public are equally at risk to exposure to *Legionella* bacteria, and if the business is involved in the following there is a greater risk:

- leisure services
- hotels

- care homes and healthcare
- residential homes.

Having established whether there is a water system present that could potentially become contaminated with *Legionella*, the second step is to establish the likelihood of exposure to the hazard (i.e. the risk).

Consider whether the water system:

- operates below 20°C
- operates above 45°C
- operates somewhere in between 20°C and 45°C
- is regularly disinfected
- contains or generates water in droplet, spray or mist form
- is regularly cleaned.

Remember, as with all bacteria, *Legionella* need the opportunity to multiply to be infective, and if the conditions that are ideal for this are prevented, the risk will be reduced significantly.

Step three will be to establish that there is the potential risk of Legionnaires' disease being spread and to determine the control measures that will eliminate the hazard or reduce it to acceptable levels.

The risk assessment must be in writing and reviewed regularly.

Are there any legally laid down requirements for the format of risk assessments?

No. Risk assessments can be in any format as long as they contain the necessary information to identify the hazards, the risks from the hazards, those exposed or affected by the hazards, and the control measures necessary to eliminate or reduce the risks to an acceptable level.

What are some control measures that I can put in place to reduce the risks from *Legionella* bacteria?

The following are a range of common sense control measures for all water systems that have the potential to become contaminated with *Legionella* bacteria and spread Legionnaires' disease. Suitable precautions will effectively reduce the risk of an outbreak occurring.

Hot- and cold-water services
- Keep water temperatures *below* 20°C or *above* 45°C
- fit thermostatic tap valves

- store hot water at 60°C or above
- circulate water at 50°C
- keep pipes well insulated and avoid cold-water pipes being affected by hot pipes
- keep pipe runs short
- avoid 'dead legs'
- run water from all outlets regularly – flush the system through.

Cooling towers
- Notify the cooling tower to the local authority
- clean and disinfect the system at least every 6 months
- treat the water to prevent scale, algae and microbiological build up
- take water samples for analysis
- follow manufacturers' instructions
- fit drift eliminators
- replace water-cooled systems with air-cooled systems
- use 'dip slides' to monitor microbiological activity weekly.

Other water systems
- Clean and disinfect equipment regularly
- descale shower heads – bacteria can survive on the scale
- keep temperatures as for hot and cold water
- disinfect water systems
- backwash and clean filtration systems weekly.

How often should I test water for *Legionella*?

It depends on the system that you have and the outcome of your risk assessment. For open systems, such as cooling towers, evaporative condensers and spa pools, etc. routine testing should be carried out at least every 3 months. However, there may be circumstances where more frequent sampling is required.

For hot- and cold-water systems, which are generally enclosed (i.e. not open to the elements) and not liable to significant contamination in the same way as cooling towers, microbiological monitoring is not usually required. But there may be circumstances where testing for *Legionella* is necessary (e.g. where there is doubt about the efficacy of the control regime, or where recommended temperatures or disinfection concentrations are not being consistently achieved).

Is it necessary to clean and disinfect my water system?

It is important to maintain the cleanliness of your water system. The mechanisms and frequency for doing this will depend on the system you have and whether cleaning or disinfecting is being done routinely or because of a problem identified during monitoring.

The frequency and method of routine cleaning and disinfecting should be identified in your risk assessment. This will take account of factors such as whether the system is open or closed, the type and level of contamination, and the population that could potentially be exposed.

Do all hot- and cold-water systems need an assessment, even lower risk systems?

All systems require a risk assessment, although not all systems will require elaborate control measures. A simple risk assessment may show that the risks are low and being managed properly to comply with the law. In such cases, you may not need to take further action, but it is important to review your assessment regularly in case of any changes in the system, and specifically if there is reason to suspect the assessment is no longer valid.

An example of a low-risk situation may be found:

- in a small building without individuals especially 'at risk' from *Legionella* bacteria
- where daily water usage is inevitable and there is sufficient turnover in the entire system
- where cold water is directly from a wholesome mains supply (no water-storage tanks)
- where hot water is fed from instantaneous heaters or low-volume water heaters (supplying outlets at 50°C)
- where the only outlets are toilets and hand-washing basins (no showers).

I have a number of thermostatic mixing valves in my premises. What are the key issues?

Thermostatic mixing valves (TMVs) are an important mechanism to prevent scalding in many buildings such as health and social care settings, residential buildings and public buildings where, for a number of reasons, there are increased water temperatures. A TMV will reduce the water temperature at the outlets to prevent scalding, but the potential scalding risk should be assessed and controlled in the context of the vulnerability of those using the facilities. The approach will depend on the needs and capabilities of the public, patients or residents. Where vulnerable people are identified and have access to showers or baths and the scalding risk is considered significant, type 3 TMVs are required.

The reduction in the temperature of the outlet has the potential to increase risks of *Legionella* bacteria growth. To manage this risk, the TMVs should be sited as close as possible to the point of use and flushed regularly.

TMVs should be clearly identified in the risk assessment, and the specific measures needed to ensure that risks are controlled must be included.

What are the duties in respect of keeping written records of our management procedures for *Legionella*?

If you have five or more employees you have to record any significant findings, including employees identified as being particularly at risk and the steps taken to prevent or control risks. If you have fewer than five employees you do not need to write anything down, although it is useful to keep a written record of what you have done.

Records should include details of the:

1 person or persons responsible for conducting the risk assessment and managing and implementing the written scheme
2 significant findings of the risk assessment
3 written control scheme and details of its implementation
4 details of the state of operation of the system (i.e. in use or not in use)
5 results of any monitoring inspection, test or check carried out, and the dates on which they were done.

These records should be retained throughout the period for which they remain current and for at least 2 years after that period. Records kept in accordance with (5) should be retained for at least 5 years.

It will also be good practice to keep records of all water-temperature checks, flushing frequencies, pipe and system maintenance, and periodic testing and servicing of boilers and pipework systems.

What happens if an outbreak of Legionnaires' disease occurs?

A full investigation is carried out by the local control of infection team, which usually comprises members from environmental health, the Health and Safety Executive, public health, the communicable diseases unit, the health trust and, if fatalities have occurred, the police.

The need to identify the source of the infection is paramount – hence the requirement to notify the local authority of the existence of water-cooling towers.

Enforcement officers have the powers to close buildings, prohibit the use of equipment or plant, etc. if they suspect that there is an 'imminent risk' to a person's safety.

Officers will visit premises and check records of maintenance, cleaning and bacteriological sampling. The records of water temperature are important.

Questionnaires will probably be sent to people who are known to have been affected or who have been in the vicinity of suspect buildings or in given areas.

Legionella bacteria can drift for considerable distances on the wind, so the source of the infection could be several hundred metres away from where people appeared to contract the disease.

Managers, employers, water service companies, etc. involved with the suspect plant or system could be interviewed under Section 20 of the Health and Safety at Work etc. Act 1974, or under caution under the Police and Criminal Evidence Act 1984.

What should an employer do if bacteriological results come back positive for *Legionella*?

Unless the employer is competent to interpret the results, they must consult and seek advice from experts so that the severity of risk can be determined.

The laboratory undertaking the analysis will be able to help with practical advice.

In general, if *Legionella* are identified it will indicate that something has gone wrong with the maintenance and cleaning regime. Shut down any system so that you reduce the risk of the contamination spreading.

Drain down any water system, clean and disinfect all aspects of tanks, pipes, shower heads, taps, etc.

Seek advice from competent water-treatment companies.

Should a water risk assessment be considered for other hazards associated with water?

Yes. Water can be a great hazard and cause death and injury if the risks are not fully identified and control measures implemented.

Water risk assessments should be completed whenever a hazard is identified, such as:

- risk of drowning
- risk of infection (e.g. from Weil's disease)
- surges of water
- unknown depth of water
- currents and rip tides
- overcrowding of the water facility (e.g. in swimming pools).

The risk assessment will be approached in the same way as for all risk assessments:

- identify the hazards
- identify the risks (i.e. the likelihood of harm)
- identify who could be harmed
- determine control measures to reduce the severity of the risk
- determine the need for training and sharing the information.

Drowning or contracting a water-borne disease

The employer has a duty to assess the risk. For example, how could an employee drown, how likely is it, and what work activity could be changed to reduce the risk? The outcome of the assessment must be written down.

Does a water risk assessment need to be carried out for a swimming pool?

Yes. In addition to a general risk assessment, a specific water risk assessment should be undertaken with regard to the water quality and safety. This will help determine what control measures may need to be implemented to manage any risks from bacteriological contaminants, uneven depth issues, slips and trips, etc.

Will bathers be at risk of drowning because of unusual or unexpected depth changes, are there risks associated with water inlets and drainage outlets?

Will there be risks from over- or under-dosing of disinfecting chemicals, unexpected discharges, etc?

What is the user profile of the pool? Will people with disabilities swim, and, if so, what emergency procedures will be in place?

All these eventualities need to be considered and managed where residual hazards and risks are high or unacceptable.

Should a specific risk assessment be carried out for a spa pool or hot-tub?

Yes. Spas and hot tubs, whirlpool baths and Jacuzzis can all be major sources of *Legionella* bacteria and a cause of Legionnaires' disease.

If you are responsible for managing spa baths you need to identify and assess any potential sources of *Legionella*, and consider who and how people could be exposed: in other words, conduct a risk assessment.

When conducting the risk assessment the manager or responsible person must consider the individual nature of the premises and spa pool(s). To help achieve this it is important that an up-to-date schematic diagram is kept of the spa pool(s) and associated plant. This can be used to decide which parts of the system pose a risk to workers and users.

- Prepare a plan to prevent or control any risks you have identified.
- Implement, manage and monitor the precautions you put in place.
- Keep records of this work.
- Appoint someone to manage this responsibility if you cannot do it yourself.
- Train staff to operate the spa bath correctly, giving them appropriate information about the risks and the plan for managing the risks.

The person who conducts the risk assessment should have adequate knowledge, training and expertise to understand the hazard (i.e. the presence of infectious agents in the spa pool) and the risk.

They should know how running the spa pool produces the hazard, have the ability and authority to collect all the information needed to do the assessment, and have the knowledge, skills and experience to make the right decisions about the risks and the precautions needed.

The following general factors need to be considered when carrying out the risk assessment:

- the source of the supply water (e.g. mains supply or an alternative)
- possible sources of contamination of the water supply (e.g. biofilms within the pipework, bathers, soil, grass, leaves (the latter for spa pools sited outdoors))
- the normal operating characteristics of the spa pool
- the people who will be working on or near the spa pool or using it
- the measures chosen to adequately control exposure, including the use of personal protective equipment
- unusual, but reasonably foreseeable, operating conditions (e.g. breakdowns)
- the type, design, size, approximate water capacity and designed bather load of the spa pool
- the type of dosing equipment, including the use of automatic controls, pump arrangements, balance tanks, air blowers, etc.
- the piping arrangements and construction materials
- the type of filtration system
- the heat source and design temperature
- the chemical dosing equipment, including chemical separation, personal protective equipment and chemical storage arrangements (e.g. bunding)

- the type of treatment used to control microbiological activity (e.g. chlorine)
- the method used to control the water pH (e.g. sodium bisulphate)
- the cleaning regime, including the ease of cleaning, what is cleaned, and how and when
- the testing regime, including microbiological tests, the frequency of tests, operating parameters and the action required when the results are outside the set parameters.

The significant findings of the risk assessment should be recorded. It will be easier to demonstrate to enforcing authorities that a suitable risk assessment has been done if there is a written version. It is essential that the effectiveness of control measures is monitored.

The risk assessment should be reviewed regularly (at least every 2 years) and whenever there is reason to suspect it is no longer valid, for example when:

- there are changes to the spa pool or the way it is used
- there are changes to the premises the spa pool is installed in
- new information is available about the risks or control measures
- the results of tests indicate control measures are not effective
- an outbreak of a disease (e.g. Legionnaires' disease) is associated with the spa pool.

Case study

NHS trust prosecuted for failing to act on positive *Legionella* results and failing to manage *Legionella*

Brighton and Sussex University Hospitals NHS Foundation Trust was fined £50 000 for failing to control *Legionella*.

The trust, which runs the Royal Sussex County Hospital in Brighton, was sentenced after a joint investigation by the Health and Safety Executive (HSE) and Sussex Police identified a history of failing to manage the deadly water-borne bug.

The investigation followed the death of a vulnerable cancer patient at the Royal Sussex on 9 November 2011 – 8 days after a urine tested positive for the *Legionella* bacteria antigen.

An inquest found that the patient died of natural causes, and that by the time of her death the *Legionella pneumonia* appeared to have been successfully treated. However, the inquest found that the infection may have hastened the patient's death.

The court heard that, although the trust was monitoring *Legionella* and water temperatures across its various sites at the time of the patient's death, between

October 2010 and November 2011 a total of 114 positive *Legionella* tests and a further 651 records of water temperatures outside the required parameters were not adequately acted upon.

Chlorine dioxide units were fitted at five sites to control the bacteria, but HSE inspectors found that the units routinely failed to emit the required dosage to work effectively. High *Legionella* readings were detected in at least five buildings.

Inspectors also found that hot water often failed to reach the 60°C temperature needed to kill the *Legionella*, this being another control system that the trust relied on.

The court was told that one of the major contributors to the serious control failures was the fact that staff did not have sufficient instruction, training and supervision to be able to make informed decisions and take appropriate action.

The intervention of the HSE and Sussex Police after the patient's death resulted in the trust instigating a new management system to control *Legionella* effectively.

Brighton and Sussex University Hospitals NHS Foundation Trust was fined £50 000 and ordered to pay costs of £38 705.60 after pleading guilty to breaching Section 3(1) of the Health and Safety at Work etc. Act 1974.

After sentencing, HSE Inspector Michelle Canning commented:

> The *Legionella* control failures we identified at the Royal Sussex are made all the more stark by the fact that those most at risk of contracting *Legionella* were among the most vulnerable in our society – including cancer patients.

> All organisations have a legal duty to control the risks arising from hot- and cold-water systems, but healthcare providers like hospital trusts must be especially vigilant and robust in terms of the systems they have in place.

Risk Assessments: Questions and Answers
ISBN 978-0-7277-6076-0

ICE Publishing: All rights reserved
http://dx.doi.org/10.1680/raqa.60760.269

Chapter 17
Fire risk assessments

What is the law that governs fire safety in the workplace?

The Regulatory Reform (Fire Safety) Order 2005 governs fire safety in the workplace and sets out the requirements for employers and others to manage fire safety for employees and others affected by their business.

What is the legal requirement for a fire risk assessment?

The Regulatory Reform (Fire Safety) Order 2005 requires the 'responsible person' to make a suitable and sufficient assessment of the risks to which relevant persons are exposed in respect of fire.

A fire risk assessment must be undertaken of any premises in which persons are employed or to which others have access.

The principles of risk assessment to be followed are the same as those as listed in the Management of Health and Safety at Work Regulations 1999.

Where there are five or more employees the significant findings of the risk assessment must be recorded in writing.

What type of premises does the Regulatory Reform (Fire Safety) Order 2005 apply to?

The order applies to virtually all premises and covers nearly every type of building, including open spaces.

The order applies to:

- offices
- shops
- factories
- care homes
- hospitals
- places of worship

- village and community halls
- schools, colleges and universities
- pubs, clubs and restaurants
- sports centres
- tents and marquees
- hotels and hostels
- warehouses
- shopping centres.

It does not apply to private dwellings, including individual flats in a block or a converted house.

(NB: Fire safety requirements for houses in multiple occupation are contained in the legislation on housing.)

Who is legally responsible for complying with the Regulatory Reform (Fire Safety) Order 2005?

The order refers to a 'responsible person' as being the person with responsibility for complying with the order.

In the workplace, the responsible person will be the employer or any other person who may have control of any part of the premises (e.g. owner or occupier).

In all other premises the person or persons in control of the premises will be responsible for complying with the order.

What happens if there is more than one person in control of the premises?

Where there is more than one responsible person in any type of premises (e.g. multi-occupied complex), they all must take reasonable steps to cooperate with one another.

If there are managing agents responsible for common parts of a premises then they will be the responsible person for those areas.

The requirements of the order are in fact imposed on *any* person who has, to any extent, control of premises insofar as the requirements of the order relate to matters within their control.

What are the main requirements or rules under the Regulatory Reform (Fire Safety) Order 2005?

The responsible person must:

- carry out a fire risk assessment to identify any possible dangers and risks of fire
- consider who may be especially at risk
- reduce the risk of fire within the premises as far as is reasonably practicable
- provide general fire precautions to deal with any residual fire risk
- implement special measures to control the risks from flammable materials, explosive materials or other hazardous substances
- create an emergency plan for fire safety
- record significant findings in writing
- review and monitor the fire safety arrangements.

In addition, the responsible person must ensure everybody who could be affected by the business undertaking receives information or training and instruction.

The responsible person must also prepare an emergency plan for fire safety arrangements.

Any equipment or facilities (e.g. fire-fighting lifts) provided for the use and safety of fire-fighters must be adequately maintained.

Does the employer's duty to manage fire safety apply only to employees?

No. The order refers to all 'relevant persons', which the order defines as:

(a) any person (including the responsible person) who is or may be lawfully on the premises; and
(b) any person in the immediate vicinity of the premises who is at risk from a fire on the premises.

This includes everyone – employees, public, customers, contractors, visitors, students, patients, etc.

The responsible person must consider the fire safety needs of those at special risk, such as disabled people, young people and people not familiar with the building.

The Regulatory Reform (Fire Safety) Order 2005 introduces a clear responsibility for the 'responsible person' to consider the needs of the public in respect of fire safety, not just employees.

What are the main requirements of the Regulatory Reform (Fire Safety) Order 2005?

The main requirement of the order is for the responsible person to carry out a fire risk assessment.

The fire risk assessment must focus on the safety in cases of fire of all 'relevant persons'.

In addition to a fire risk assessment, the responsible person must:

- appoint one or more competent persons to assist with fire safety
- provide employees with clear and relevant information on the risks to them identified by the fire risk assessment, about measures taken to prevent fires, and generally how they will be protected in the event that a fire occurs
- consult with employees about fire safety matters
- take special steps regarding the employment of young people and inform their parents about the fire measures that have been taken
- inform all persons who are not employees (e.g. contractors and temporary workers) about fire safety measures in place in the premises
- cooperate with other responsible persons and coordinate fire safety matters in multi-occupied buildings
- provide the employer of any temporary workers from an outside organisation with clear and relevant information about how the safety of their employees will be protected in the event of fire
- consider any special precautions needed for managing fire safety with respect to any dangerous, flammable or hazardous substances, including explosives
- establish a suitable means for contacting the emergency services and providing them with relevant information about dangerous substances
- provide all relevant persons, where reasonable, with suitable information, instruction and training on fire safety
- provide and maintain suitable fire-fighting equipment, fire-detection equipment, fire warning systems, emergency lighting, etc. as is deemed necessary in the premises by the fire risk assessment
- advise employees that they have a duty to cooperate with their employer so as to enable the employer to comply with their statutory duties.

Can anyone be appointed as the competent person?

Yes, provided they have enough knowledge and experience to understand the basic principles of fire safety and to be able to implement the measures contained in the Regulatory Reform (Fire Safety) Order 2005.

No formal qualifications are required to become a competent person but, as the onus is on the responsible person to appoint someone – or more than one person if necessary – to provide safety assistance, the responsible person will need to be able to demonstrate that the person so appointed has suitable credentials to take on the role.

The type of people who could be classed as competent persons could be:

- fire safety consultants
- fire engineers
- health and safety managers or officers – provided they have undergone fire safety training
- employers – if they have attended basic fire safety training
- unit or department managers – provided they have undergone fire safety training
- health and safety consultants.

The competent person is there to help the responsible person fulfil their statutory duties.

What is a fire risk assessment?

A fire risk assessment is, in effect, an audit of the workplace and work activities in order to establish how likely a fire is to start, where it would be, how severe it might be, who it would affect and how people would get out of the building in an emergency.

A fire risk assessment should be concerned with *life safety*, not with matters that are really fire engineering matters.

A fire risk assessment is a structured way of looking at the hazards and risks associated with fire and the products of fire (e.g. smoke).

Like all risk assessments, a fire risk assessment follows *five key steps*, namely:

1 identify the hazards
2 identify the people and the location of people at significant risk from a fire
3 evaluate the risks
4 record the findings and actions taken
5 keep the assessment under review.

So, a fire risk assessment is a record that shows you have assessed the likelihood of a fire occurring in your workplace, identified who could be harmed and how, decided on what steps you need to take to reduce the likelihood of a fire (and therefore its harmful consequences) occurring, and recorded all these findings regarding your premises in a particular format called a 'risk assessment'.

Definitions

Risk assessment – the overall process of estimating the magnitude of risk and deciding whether or not the risk is tolerable or acceptable.

Risk – the combination of the likelihood and consequence of a specified hazardous event occurring.

> **Hazard** – a source or a situation with a potential to harm in terms of human injury or ill health, damage to property, damage to the environment or a combination of these.
>
> **Hazard identification** – the process of recognising that a hazard exists and defining its characteristics.

What are the five steps to undertaking a fire risk assessment?

Step 1: Identify the hazards

(a) Sources of ignition

Identify the sources of ignition in your premises by looking for possible sources of heat that could get hot enough to ignite the material in the vicinity.

Such sources of heat/ignition could be:

- naked flames (e.g. candles, fires and blow lamps)
- electric, gas or oil-fired heaters
- hot-work processes (e.g. welding and gas cutting)
- cooking, especially frying
- faulty or misused electrical appliances, including plugs and extension leads
- lighting equipment, especially halogen lamps
- hot surfaces and obstructions of ventilation grills (e.g. radiators)
- poorly maintained equipment that causes friction or sparks
- static electricity
- arson
- smokers' material, especially from any illicit smoking in prohibited areas.

Look out for any evidence that things or items have suffered scorching or overheating (e.g. blackened plugs and sockets, burn marks, cigarette burns and scorch marks).

Check each area of the premises systematically:

- customer area, public areas and reception
- work areas and offices
- staff kitchen and staff rooms
- store rooms and cleaners' stores
- plant rooms and motor rooms
- refuse areas
- external areas.

(b) Sources of fuel

Anything (generally) that burns is fuel for a fire. Fuel can also be invisible in the form of vapours, fumes, etc. given off from other less flammable materials.

Look for anything in the premises that is present in sufficient quantity to burn reasonably easily, or to cause a fire to spread to more easily flammable fuels.

Fuels to look out for are:

- wood, paper and cardboard
- flammable chemicals (e.g. cleaning materials)
- flammable liquids (e.g. cleaning substances and liquid petroleum gas)
- flammable liquids and solvents (e.g. white spirit, petrol and methylated spirit)
- paints, varnishes, thinners, etc.
- furniture, fixtures and fittings
- textiles
- ceiling tiles and polystyrene products
- waste materials and general rubbish
- gases.

Consider also the construction of the premises – are there any materials used that would burn more easily than other types. Hardboard, chipboard and blockboard burn more easily than plasterboard.

(c) Identifying sources of oxygen

Oxygen is all round us in the air that we breathe. Sometimes, other sources of oxygen are present that accelerate the speed at which a fire ignites (e.g. oxygen cylinders for welding).

The more turbulent the air the more likely the spread of fire will be. For example, opening doors brings a 'whoosh' of air into a room and the fire is fanned and intensifies. Mechanical ventilation also moves air around in greater volumes and more quickly.

Do not forget that, while ventilation systems move oxygen around at greater volumes, they will also transports smoke and toxic fumes around the building.

Step 2: Identify who could be harmed

You need to identify who will be at risk from a fire and where they will be when a fire starts. The law requires you to ensure the safety of your staff and others (e.g. customers). Would anyone be affected by a fire in an area that is isolated? Could everyone respond to an alarm, or evacuate?

Will you have people with disabilities on the premises (e.g. those in wheelchairs, or visually or hearing impaired)? Will they be at any greater risk of being harmed by a fire than other people?

Will contractors working in plant rooms, on the roof, etc. be adversely affected by a fire? Could they be trapped or not hear alarms?

Who might be affected by smoke travelling through the building? Smoke often contains toxic fumes.

Step 3: Evaluate the risks arising from the hazards

What will happen if there is a fire? Does it matter whether it is a minor or major fire? Remember that small fires can grow rapidly into infernos.

A fire is often likely to start because:

- people are careless with cigarettes and matches
- people purposely set light to things
- cooking canopies catch fire due to a build-up of grease
- people put combustible material near flames or ignition sources
- equipment is faulty because it is not maintained
- electrical sockets are overloaded.

Will people die in a fire from:

- flames
- heat
- smoke
- toxic fumes?

Will people get trapped in the building?

Will people know that there is a fire, and will they be able to get out?

This step of the risk assessment is about looking at what *control measures* you have in place to help control the risk or reduce the risk of harm from a fire.

Remember – fire safety is about *life safety*. Get people out fast and protect their lives. Property is always replaceable.

You will need to record on your fire risk assessment the fire precautions you have in place.

- What emergency exits do you have, and are they adequate and in the correct place?

- Are they easily identified and unobstructed?
- Is there fire-fighting equipment?
- How is the fire alarm raised?
- Where do people go when they leave the building – an assembly point?
- Are the signs for fire safety adequate?
- Who will check the building and take charge of an incident (i.e. do you have a fire warden appointed)?
- Are fire doors kept closed?
- Are ignition sources controlled and fuel sources managed?
- Do you have procedures to manage contractors? (Remember: Windsor Castle went up in flames because a contractor used a blow torch near the curtains!)

Consider all your fire safety precautions for the premises, is there anything more that you need to do?

Are staff trained in what to do in an emergency? Can they use fire extinguishers? Do you have fire drills? Is equipment (e.g. emergency lights and fire alarm bells) serviced and checked?

Step 4. Record findings and action taken
Complete a fire risk assessment form and keep it safe.

Make sure that you share the information with staff.

If contractors come to the site, make sure that you discuss *their* fire safety plans with them and that you tell them what your fire precaution procedures are.

Step 5: Keep the assessment under review
A fire risk assessment needs to be reviewed regularly – about every 6 months or so, and whenever something has changed, including a new layout, new employees, new procedures, new legislation, increased stock, etc.

What are the principles of prevention in respect of fire safety?
The Regulatory Reform (Fire Safety) Order 2005 requires all responsible persons to undertake fire risk assessments of their premises, operations and any other business activity that could expose persons to risks to their safety.

If the risk assessment indicates that preventive and protective measures are required, these measures must be in accordance with the requirements of Schedule 1, Part 3 of the Regulatory Reform (Fire Safety) Order 2005.

The principles of prevention are:

- avoiding risks
- evaluating the risks that cannot be avoided
- combating risks at source
- adapting to technical progress
- replacing the dangerous with the non-dangerous or less dangerous
- developing a coherent overall protection policy that covers technology, organisation of work and the influence of factors relating to the working environment
- giving collective protective measures priority over individual protective measures
- giving appropriate instructions to employees.

The principles of prevention listed above are more detailed than in other health and safety legislation.

The fire officer has indicated that my fire risk assessment is not 'suitable and sufficient'. What should I do about it?

If possible, discuss it with the fire officer. Sometimes it is the format of a risk assessment that the fire risk officer does not like rather than its content.

The fire officer may be concerned that you have not considered all aspects of fire safety and that your risk assessment is missing a vital element. Generally, the fire officer should tell you why they are not satisfied with your risk assessment.

If you feel that you have covered everything and that the fire officer is being overzealous, ask to discuss the matter with a senior officer.

Be prepared and review your risk assessment together with your competent person, or seek professional advice.

A fire risk assessment must address the major fire safety hazards associated with your workplace – it does not have to be perfect.

Does a fire risk assessment need to be carried out every year?

No, not necessarily. The responsible person has a duty to ensure that the fire risk assessment is kept up to date and is relevant to the risk of fire within the premises.

A fire risk assessment should be reviewed regularly and a record of the review kept with the master fire risk assessment.

Whenever there are changes in circumstances, such as building works, new equipment, furniture layout changes, increases in staff numbers and so on, a reassessment should be carried out to see if additional fire safety controls are required or if further training is needed.

Is there any guidance on assessing the risk rating of premises in respect of fire safety?

When completing fire risk assessments it is sensible to categorise any *residual risk* for the premises into a risk rating category – normally 'high', 'medium' or 'low'.

In terms of fire risk rating it is usual to refer to a medium risk as 'normal'.

The Government publication *Guide to Fire Safety: Employers Guide to Fire Safety* gives some guidance on how to fire risk rate premises. Also, the Fire Safety Guides published to support the Regulatory Reform (Fire Safety) Order 2005 (https://www.gov.uk/government/collections/fire-safety-law-and-guidance-documents-for-business)   give some advice on risk rating for premises.

High-risk premises

- Any premises where highly flammable or explosive substances are stored or used (other than in very small quantities).
- Any premises where the structural elements present are unsatisfactory in respect of fire safety:
  - lack of fire-resisting separation
  - vertical or horizontal openings through which fire, heat and smoke can spread
  - long and complex escape routes created by extensive subdivision of floors by partitions, etc.
  - complex escape routes created by the positioning of shop-unit displays, machinery, etc.
  - large areas of smoke- or flame-producing furnishings and surface materials, especially on walls and ceilings.
- Permanent or temporary work activities that have the potential for fires to start and spread, for example:
  - workshops using highly flammable materials and substances
  - paint spraying
  - activities using naked flames (e.g. blow torches and welding)
  - large kitchens in work canteens or restaurants
  - refuse chambers and waste disposal areas
  - areas containing foam or foam-plastic upholstery and furniture.
- Where there is significant risk to life in case of fire:
  - sleeping accommodation provided for staff, guests, visitors, etc. in significant numbers

- treatment or care facilities where occupants have to rely on others to help them
- high proportions of elderly or infirm people
- large numbers of people with disabilities
- where people work in remote areas (e.g. plant rooms and roof areas)
- large numbers of people resort to the premises relative to its size (e.g. sales at retail shops)
- large numbers of people resorting to the premises where the number of people available to assist is limited (e.g. entertainment events and banquets).

Normal-risk premises
■ Where an outbreak of fire is likely to remain contained to localised areas or is likely to spread only slowly, allowing people to escape to a place of safety.
■ Where the number of people in the premises is small and people are likely to escape via well-defined means of escape to a place of safety without assistance.
■ Where the premises has an automatic warning system or an effective automatic fire-fighting, fire-extinguishing or fire-suppression system which may reduce the risk categorisation from 'high'.

Low-risk premises
■ Where there is minimal risk to peoples' lives and where the risk of fire occurring is low, or the potential for fire, heat or smoke spreading is negligible.

Who must I consider when preparing my risk assessment?
Responsible persons must consider the following people as being at risk in the event of a fire:

■ employees and those on temporary or agency contracts
■ employees whose mobility, sight or hearing might be impaired
■ employees with learning difficulties or mental illness
■ other persons on the premises if the premises are multi-occupied
■ anyone occupying remote areas of the premises
■ visitors and members of the public, including contractors, etc.
■ anyone who may sleep on the premises
■ anyone with any special needs or disabilities.

Does a fire risk assessment have to consider members of the public?
A fire risk assessment must be carried out by the responsible person and must consider the risks to the safety of *relevant persons* (i.e. all persons who are, or could be, lawfully on the premises). This will include members of the public.

The Regulatory Reform (Fire Safety) Order 2005 has made the inclusion of all persons, including the public, a legal requirement when completing risk assessments.

Who can carry out a fire risk assessment?
The Regulatory Reform (Fire Safety) Order 2005 states that the person who carries out a fire risk assessment shall be *competent* to do so. They do not necessarily have to have had formal training.

Competency is not defined specifically in the order but is generally taken to mean having a level of knowledge and experience that is relevant to the task in hand.

A fire risk assessment is a logical, practical review of the likelihood of a fire starting on the premises and the consequences of such a fire. Someone who has good knowledge of the work activities and the layout of the building, together with some knowledge of what causes a fire, would be the person best placed to carry out a fire risk assessment.

What fire hazards need to be considered?
Consider any significant fire hazards in the room or area under review.

- combustible materials (e.g. large quantities of paper, or combustible fabrics and plastics)
- flammable substances (e.g. paints, thinners, chemicals, flammable gases and aerosol cans)
- ignition sources (e.g. naked flames, sparks, portable heaters, smoking materials and hot-works equipment).

Do not forget to consider materials that might smoulder and produce quantities of smoke. Also, consider anything that might give off toxic fumes.

Consider also the type of insulation involved or used in cavities, roof voids, etc. Combustible material may not always be visible (e.g. hidden cables in wall cavities).

What structural features are important to consider when carrying out a fire risk assessment?
Fire, smoke, heat and fumes can travel rapidly through a building if not restricted by fire protection and compartmentalisation.

Any part of a building that has open areas, open staircases, etc. will be more vulnerable to the risk of fire should one start.

Openings in walls and large voids above ceilings and below floors allow a fire to spread rapidly. Large voids also usually contribute extra ventilation, thereby adding more oxygen to the fire.

One method of fire prevention is to use fire-resistant materials and to design buildings so that fire will not travel from one area to another.

Any opportunity for a fire to spread through the building must be noted on the fire risk assessment.

What are some of the factors to consider when assessing existing control measures for managing fire safety?

Many premises and employers already have some level of fire safety management in place, and the Regulatory Reform (Fire Safety) Order 2005 is not intended to add a heavy burden on employers and other responsible persons.

Existing control measures must be reviewed. The following are examples of what to look for:

- the likely spread of fire
- the likelihood of a fire starting
- the number of occupiers of the area
- the use of the premises and the activity undertaken
- the time available for escape
- the means of escape
- the clarity of the escape
- the effectiveness of signage
- how the fire alarm is raised
- whether the fire alarm can be heard by everyone
- travel distances to exits
- the number and widths of exits
- the condition of corridors
- storage and obstructions
- inner rooms and dead ends
- the type of and access to staircases
- openings, voids, etc. within the building
- the type of fire doors
- the use of panic bolts
- unobstructed fire doors
- intumescent strips
- well-fitting fire doors

- propped open fire doors
- the type of fire alarm
- the location, number and condition of fire extinguishers
- the display of fire safety notices
- emergency lighting
- maintenance and testing of fire alarm break glass points
- installation of sprinklers
- the location and condition of smoke detectors
- the use of heat detectors
- adequate lighting in an evacuation
- training of employees
- practised fire drills
- general housekeeping
- management of contractors
- use of hot-works permits
- control of smoking
- fire safety checks
- provision for managing the safety of people with disabilities
- special conditions (e.g. storage of flammable substances)
- storage of combustible materials near a heat source.

The best fire risk assessments are 'site specific' – review and inspect your *own* workplace. An example template for a fire risk assessment is given at the end of this chapter.

Where can I obtain further information on complying with the Regulatory Reform (Fire Safety) Order 2005?

All fire authorities should be able to give you information and advice on complying with fire safety requirements.

However, in order to address the needs of the business community in understanding the duties placed on them by the Regulatory Reform (Fire Safety) Order 2005, the Government has published a series of fire safety risk assessment guides. These guides set out everything an employer – or responsible person – will need to do to comply with the law and to ensure that fire safety standards are maintained.

The guides cover:

- offices and shops
- factories and warehouses
- sleeping accommodation
- small and medium-sized places of assembly

- large places of assembly
- theatres and cinemas
- educational premises
- residential care premises
- outdoor events
- healthcare premises
- transport premises and facilities.

All the guides are available for free download on the government website: https://www.gov.uk/government/collections/fire-safety-law-and-guidance-documents-for-business.

FIRE RISK ASSESSMENT AND EMERGENCY PLAN – EXAMPLE FOR HOSPITALITY PREMISES

FIRE SAFETY RISK ASSESSMENT AND EMERGENCY PLAN FOR

<insert name and address of premises>

Prepared by: *<insert name>*

Date: *<insert date>*

CONTENTS

1.0 INTRODUCTION

2.0 FIRE SAFETY POLICY STATEMENT

3.0 PREMISES DETAILS

4.0 SIGNIFICANT FINDINGS OF FIRE RISK ASSESSMENT

5.0 FIRE HAZARDS

    5.1 Kitchen

    5.2 Bars/restaurants/serveries

    5.3 Customer areas, lounges and front of house areas, including meeting rooms

    5.4 Sleeping accommodation (including staff accommodation, if applicable)

    5.5 Back of house areas/storerooms

    5.6 Plant rooms/equipment rooms

    5.7 External areas (consider arson possibilities)

    5.8 Other areas (please describe)

6.0 PERSONS AT RISK

7.0 FIRE SAFETY CONTROL MEASURES

    7.1 Fire safety management

    7.2 Fire safety training

    7.3 Means of escape

1.0   INTRODUCTION

This Fire Risk Assessment has been completed as required by the Regulatory Reform (Fire Safety) Order 2005 and has been formulated following guidance from a number of fire authorities.

It consists of a structured and systematic examination of the workplace to identify the hazards from fire and the control measures implemented to reduce the risks to life safety.

Recommended further actions required to reduce further the risks associated with fire are detailed in Section 4.0, Significant Findings of Fire Risk Assessment.

Fire risk assessment review recommendation

This Fire Risk Assessment must be reviewed when there is any significant change to the premises' normal activities (e.g. when any major refurbishment takes place) or at the end of 12 months from initial completion (i.e. *<insert date 12 months from date of initial Fire Risk Assessment>*).

A fire risk assessment is a dynamic process, and fire safety matters must be considered in all day-to-day activities.

2.0 FIRE SAFETY POLICY STATEMENT

<Insert name of person acting on behalf of the employer> is deemed to be the responsible person for fire safety as required under the Regulatory Reform (Fire Safety) Order 2005 (RRO), and as such they have reviewed all existing fire safety arrangements and implemented improvements as appropriate across all premises.

The responsibility for fire safety management in each premises is delegated to the General Manager. Each General Manager is regarded as a competent person for managing fire safety, as required by the RRO. They have received suitable training in fire safety awareness and are aware of the management procedures necessary to protect from fire staff, customers and others who may be using the facilities.

Every premises has a Fire Risk Assessment, Emergency Action Plan, Fire Safety Management Plan and Fire Risk Assessment for Employees and Visitors with Disabilities, where appropriate.

Fire risk assessments will be reviewed regularly by the premises manager and amended as necessary. The Fire Risk Assessment will be formally reviewed annually by Perry Scott Nash Associates Ltd, and, where necessary, recommendations will be made to the company for amendments, improvements, etc.

All employees of *<insert name of person acting on behalf of the employer>* will receive suitable and sufficient training in fire safety awareness and emergency and evacuation procedures. This training will be regularly updated and employees' knowledge tested via regular fire safety drills and practice evacuations. All building evacuations will be classified as opportunities for review of procedures.

The Board of Directors of *<insert name of business>* is committed to ensuring the highest standards of fire safety within the premises and compliance with all relevant legislation.

Signed: . . . .   Date: . . . .

Operations Director

<insert name of business>

3.0   PREMISES DETAILS
Name and address of premises

Premises use

Telephone number

Name of Competent Person for Fire Safety (the General Manager)

Company name and address (the Responsible Persons)

General description of premises

Number of floors

```
```

Number of bedrooms (if applicable)

```
```

Number of staircases

- Number of accommodation staircases
- Number of escape staircases (protected)

```
```

Number of meeting rooms

```
```

Number of restaurants

```
```

Number of bars

```
```

Relevant Persons

- Number of employees on duty at any one time: . . . . .
- Number of day customers (approx.): . . . . .
- Number of night-time residents (approx.): . . . . .
- Number of people attending functions/meetings, etc. (max. possible occupancy): . . . . .
- Number of disabled persons or others at specific risk from fire hazards: . . . . .
- Number of young persons/children at specific risk from fire hazards: . . . . .

History of fire alarms, false alarms, etc.

```



```

RISK RATING OF PREMISES WITH CURRENT CONTROL MEASURES IN PLACE

High ☐   Medium ☐   Low ☐

RISK RATING OF PREMISES WITH ADDITIONAL CONTROLS IMPLEMENTED AS ITEMISED IN SIGNIFICANT FINDINGS

High ☐   Medium ☐   Low ☐

4.0 SIGNIFICANT FINDINGS OF FIRE RISK ASSESSMENT

Description of unsatisfactory condition	Persons at risk	Proposed remedial action	Priority	By when	By whom

Priority ratings:

High: Considered urgent action required or priority for expenditure. May not need to be completed within short timescales but considered to give greatest protection to relevant persons.

Medium: Should be completed whenever possible and as soon as possible.

Low: Considered desirable but not necessarily urgent.

5.0 FIRE HAZARDS

List the potential fire hazards in the following areas:

5.1 Kitchen

5.2 Bars/restaurants/serveries

5.3 Customer areas, lounges and front of house areas, including meeting rooms

5.4 Sleeping accommodation (including staff accommodation, if applicable)

5.5 Back of house areas/storerooms

5.6 Plant rooms/equipment rooms

5.7   External areas (consider arson possibilities)

```

```

5.8   Other areas (please describe)

```

```

6.0   PERSONS AT RISK

The following persons are deemed to be at risk from the fire hazards identified:

1   sleeping guests (if applicable)
2   customers
3   all employees
4   members of the public
5   contractors
6   people with disabilities.

However, the control measures identified in the remainder of this Fire Risk Assessment will ensure that any of the above groups of people will not be exposed to unnecessary risks to their health, safety and welfare.

7.0   FIRE SAFETY CONTROL MEASURES

7.1 Fire safety management	N/A	Yes	No
(a) Have a suitable number of employees been appointed to take charge of an event caused by fire?	☐	☐	☐
(b) Is there evidence of regular reviews of the fire safety plans?	☐	☐	☐
(c) Do visitors sign in and out and are they easily accounted for?	☐	☐	☐
(d) Does the premises have a written emergency plan for fire safety?	☐	☐	☐
(e) Has an assembly point been identified and communicated to all staff?	☐	☐	☐
(f) Are regular fire safety checks of escape routes, sources of ignition, etc. carried out by employees?	☐	☐	☐

(g) Are staff fully aware of the significant hazards of fire within the premises? □ □ □

(h) Is there confidence in the management team's competency in dealing with a major event caused by fire? □ □ □

(i) Is there a written procedure for grounding the lifts or otherwise ensuring that equipment is not used once alarms have sounded? □ □ □

(j) Is someone nominated to meet the fire brigade and to liaise with firefighters? □ □ □

7.2 Fire safety training	N/A	Yes	No
(a) Is there evidence that all members of staff have received fire safety training?	□	□	□
(b) Is there regular refresher training?	□	□	□
(c) Are suitable records kept of employee training?	□	□	□
(d) Do employees know what their duties are in the event of an emergency?	□	□	□
(e) Do staff generally know what to do in an emergency?	□	□	□
(f) Are regular fire drills carried out or does the premises use false alarms as a fire drill exercise?	□	□	□
(g) Are staff trained in how to respond to the needs of disabled persons using the premises?	□	□	□
(h) Have any young persons received more specific training on fire safety?	□	□	□
(i) Is there evidence that language barriers are overcome and that all staff comprehend the information given to them?	□	□	□

7.3 Means of escape	N/A	Yes	No
(a) Are all escape routes clear of obstructions?	□	□	□
(b) Are there sufficient fire exit signs on the escape routes?	□	□	□
(c) Are doors kept shut on escape routes?	□	□	□
(d) Are there signs indicating how to use door-opening mechanisms (e.g. 'Push bar to open')?	□	□	□
(e) Are all internal fire doors indicated by 'Fire door, keep shut' notices?	□	□	□
(f) Are internal fire-resisting doors to cupboards indicated by 'Fire door, keep locked' signs?	□	□	□
(g) Can all fire safety signs and call point signs be seen clearly?	□	□	□
(h) If doors are on magnetic locks, are there break glass overrides?	□	□	□
(i) Are the number, distribution and size of routes and exits that lead to a place of safety satisfactory?	□	□	□
(j) Are final exit doors free from fastenings (e.g. keys and locks)?	□	□	□

		N/A	Yes	No
(k)	Are travel distances in accordance with relevant guides and good practice (i.e. 18 m dead-end travel, 45 m two-way travel)?	☐	☐	☐
(l)	Are fire exits available at all material times?	☐	☐	☐
(m)	Where necessary, are there adequate safe refuge areas designated on the escape routes?	☐	☐	☐
(n)	Is there adequate means of escape for people with disabilities?	☐	☐	☐
(o)	Are external fire escape stairs in good condition and regularly maintained?	☐	☐	☐
(p)	Are any fire doors locked or not able to be opened?	☐	☐	☐
(q)	Are corridors used for any storage of combustible materials, etc.?	☐	☐	☐
(r)	If necessary, are the walls and partitions to stairways fire resisting? (Taking into account the fire hazards present, could a fire spread to the staircase before occupants have evacuated?)	☐	☐	☐

7.4	**Fire alarm and fire detection**	N/A	Yes	No
(a)	Is there a suitable audible alarm system or other means of raising the alarm in the event of fire?	☐	☐	☐
(b)	Are sufficient numbers of fire detection devices (e.g. smoke detectors and heat detectors) available within the premises?	☐	☐	☐
(c)	Does the premises have a sprinkler or suppression system?	☐	☐	☐
	■ For the whole premises?	☐	☐	☐
	■ For part of the premises?	☐	☐	☐
	■ For the kitchen?	☐	☐	☐
(d)	Is the fire alarm tested weekly and records kept?	☐	☐	☐
(e)	If a manual alarm is used, is it tested weekly and easily available?	☐	☐	☐
(f)	If a manual system is used, can the person operating the device do so in a position of safety?	☐	☐	☐
(g)	Can all occupants hear the alarm when it is sounded?	☐	☐	☐
(h)	Is the fire alarm maintained on a routine basis by a competent person?	☐	☐	☐
(i)	Are records available of the last routine maintenance check?	☐	☐	☐

7.5	**Emergency lighting**	N/A	Yes	No
(a)	Does the premises have emergency lighting installed?	☐	☐	☐
(b)	Is there sufficient illumination at changes in level?	☐	☐	☐
(c)	Is there sufficient illumination at changes in direction?	☐	☐	☐
(d)	Is there sufficient illumination to show fire alarm call points and fire-fighting equipment?	☐	☐	☐
(e)	Is it tested regularly?	☐	☐	☐

		N/A	Yes	No
(f)	Are records kept of the tests?	☐	☐	☐
(g)	Is the emergency lighting maintained on a routine basis by a competent person?	☐	☐	☐
(h)	Are records available of the last routine maintenance check?	☐	☐	☐
(i)	Is there adequate emergency lighting around external areas, on external fire escape staircases, and on any roof levels that need to be accessed for maintenance or by firefighters?	☐	☐	☐

7.6	**Fire-fighting equipment**	**N/A**	**Yes**	**No**
(a)	Are suitable fire extinguishers (e.g. water, carbon dioxide or powder) positioned around the premises, on escape routes and near to exits?	☐	☐	☐
(b)	Are extinguishers properly serviced?	☐	☐	☐
(c)	Are fire extinguishers easy to access and unobstructed?	☐	☐	☐
(d)	Are suitable signs displayed advising of use?	☐	☐	☐
(e)	Is any other fire-fighting equipment provided (e.g. wet risers and dry risers)?	☐	☐	☐
(f)	Has a competent person checked the fire extinguishers and any other fire-fighting equipment within the last 12 months?	☐	☐	☐
(g)	Are the fire extinguishers securely hung on wall brackets or placed on suitable floor plates?	☐	☐	☐

7.7	**Fire instructions**	**N/A**	**Yes**	**No**
(a)	Do all staff know what to do in the event of a fire or other emergency?	☐	☐	☐
(b)	Are Fire Action Notices displayed?	☐	☐	☐
(c)	Is an assembly point designated?	☐	☐	☐
	Where is the assembly point?	☐	☐	☐

7.8	**Housekeeping**	**N/A**	**Yes**	**No**
(a)	Is refuse removed regularly?	☐	☐	☐
(b)	Is combustible material kept away from heat sources?	☐	☐	☐
(c)	Are aerosol cans and other flammable containers stored safely?	☐	☐	☐
(d)	Are boiler rooms and electrical cupboards kept free from combustible materials?	☐	☐	☐

7.9	**Electrical**	**N/A**	**Yes**	**No**
(a)	Are sockets overloaded?	☐	☐	☐
(b)	Is the 'one plug, one socket' rule followed?	☐	☐	☐

(c) Have checks (i.e. PAT testing) been carried out on electrical plugs, leads and appliances? ☐ ☐ ☐
(d) Is equipment earthed? ☐ ☐ ☐
(e) Has the fixcd wiring been regularly checked? ☐ ☐ ☐

7.10 Smoking	N/A	Yes	No

(a) Are cigarette ends properly discarded into metal receptacles? ☐ ☐ ☐
(b) Is furnishing non-combustible? ☐ ☐ ☐
(c) Are staff trained on what to look out for with regard to smoking material and smouldering fires? ☐ ☐ ☐
(d) Do staff follow 'no smoking' rules? ☐ ☐ ☐
(e) Is there a comprehensive 'no smoking' policy in place? ☐ ☐ ☐

7.11 Heating	N/A	Yes	No

(a) If liquid pctroleum gas (LPG) heaters are used, are they serviced regularly, and subject to a risk assessment? ☐ ☐ ☐
(b) Are LPG heaters switched on by trained people? ☐ ☐ ☐
(c) Are LPG cylinders stored outside thc premises? ☐ ☐ ☐
(d) Is combustible material kept away from heaters? ☐ ☐ ☐
(e) If open fires are used, are they monitored and do they have fire guards? ☐ ☐ ☐
(f) Are heaters securely fixed? ☐ ☐ ☐

7.12 Lighting	N/A	Yes	No

(a) Are equipment, combustible materials, etc. kept away from light bulbs (i.e. light fittings)? ☐ ☐ ☐
(b) Is the correct light bulb wattage used for the lampshades? ☐ ☐ ☐

7.13 Dangerous/flammable substances	N/A	Yes	No

(a) Are substances kept away from heat sources and in metal containers? ☐ ☐ ☐
(b) Are cleaning materials correctly stored away from heat? ☐ ☐ ☐
(c) Are chemicals used safely and not near naked flames? ☐ ☐ ☐
(d) Are there any dangerous or highly flammable substances used in the premises? *If yes, please describe substances and control measures in place in the 'comments' box at the end of the section.* ☐ ☐ ☐

7.14 Storage	N/A	Yes	No

(a) Are storage areas easy to get to? ☐ ☐ ☐
(b) Are electrical sockets visible and is stock away from heat sources? ☐ ☐ ☐
(c) Are there any ignition sources? ☐ ☐ ☐

7.15 Kitchens

	N/A	Yes	No
(a) Are canopies cleaned regularly?	☐	☐	☐
(b) Are deep fat fryers thermostatically controlled?	☐	☐	☐
(c) Are grease deposits cleaned regularly?	☐	☐	☐
(d) If heat lamps/blow lamps are used for flambé dishes, is there a risk assessment?	☐	☐	☐
(e) Are thermostats working and maintained?	☐	☐	☐
(f) Is electrical and gas equipment serviced regularly?	☐	☐	☐
(g) Does gas equipment have automatic safety shut-off devices?	☐	☐	☐

7.16 Night checks (for premises with guest accommodation only)

	N/A	Yes	No
(a) Is there a proper procedure for checking the premises at night?	☐	☐	☐

7.17 Contractors

	N/A	Yes	No
(a) Are contractors' hot works controlled (e.g. permit to work procedures in place)?	☐	☐	☐
(b) Are contractors' works checked?	☐	☐	☐
(c) Is there any activity in the vicinity that could cause an increased fire risk to the premises?	☐	☐	☐
(d) Are procedures in place to advise contractors of the emergency procedures?	☐	☐	☐

7.18 Arson

	N/A	Yes	No
(a) Has appropriate consideration been given to protecting the premises, equipment, furnishings, etc. from arson?	☐	☐	☐

7.19 Special events

	N/A	Yes	No
(a) Would any special event increase the fire hazard and risk of the premises? *If yes, please describe in the 'comments' box at the end of this section.*	☐	☐	☐
(b) Has consideration been given to any legislated entertainment event?	☐	☐	☐
(c) Are candles, tea lights or other atmospheric lighting used, and if so are adequate controls in place?	☐	☐	☐
(d) Does the music equipment cut out in the event of the alarm being activated in order that customers can hear it?	☐	☐	☐

7.20 Young persons

	N/A	Yes	No
(a) Are additional procedures in place to manage fire safety in respect of young persons?	☐	☐	☐

7.21 Fire-fighters' safety	N/A	Yes	No
(a) Is there reasonable access to the premises for fire engines, etc.?	☐	☐	☐
(b) Is there anything considered likely to jeopardise the safety of firefighters other than a fire itself?	☐	☐	☐

Comments

8.0 EMERGENCY PLAN

8.1 The Duty Manager shall assume responsibility for fire safety at all times they are on duty.

The Premises Manager will assume overall responsibility for fire safety in the premises and must ensure that all legal requirements are being met.

8.2 The fire alarm will activate in the event of a suspected fire either:
- automatically via smoke/heat detectors

or
- manually via the breaking of a break glass point.

8.3 Any person who suspects that a fire has broken out can activate a break glass point.

8.4 The alarm procedure operated by the smoke/heat detectors/breaking of break glass call point will be full evacuation.

8.5 When full evacuation takes place, the Duty Manager will assume full responsibility for the evacuation and for managing fire safety.

8.6 As soon as the fire alarm has been actioned the Duty Manager must call the fire brigade by dialling 999 from the premises landline telephone or any hand-held telephone.

If the Duty Manager is unable to perform this task, the Assistant Manager must call the fire brigade. *Do not assume that someone else has called the Fire Brigade. If in doubt, call 999.*

8.7 Fire-fighting equipment is provided around the premises. The company does not expect staff to fight fires unless it is safe to do so. Small fires can spread rapidly if not quickly extinguished. The company prefers employees to raise the alarm and call the fire brigade.

However, if a small fire (e.g. in an ash bin) can be tackled safely, then the employee is to use a suitable fire extinguisher.

Red fire extinguishers containing water should be used for all fires other than:

- *fires on electrical equipment*
- *fires involving chemicals*
- *fires involving oils and grease and cooking equipment.*

8.8 The Duty Manager shall liaise with the fire brigade on its arrival. The fire brigade shall be met at the front of the premises if it is safe to do so.

Information to be relayed to the fire brigade will include:

- the name and exact address of the premises
- the location of the actual or suspected fire
- the number of persons deemed to be at risk.

8.9 On activation of the *full evacuation* alarm, the Duty Manager will ensure that:

- all lifts are grounded and deactivated (if present)
- team members are sent to check safe refuge areas, if it is safe to do so (if such areas are provided)
- team members are sent to assist in the evacuation process.
  The Duty Manager is to allocate members of staff specific duties to check areas of the premises, if it is safe to do so, as follows
- Customer toilets: *<insert position of person>*
- Kitchen: *<insert position of person>*
- Cellar: *<insert position of person>*
- Manager's office: *<insert position of person>*
- Customer areas: *<insert position of person>*

8.10 On hearing the fire alarm and on evacuation of the premises the Kitchen Manager shall turn off and shut down all gas appliances and all other equipment, as long as it is safe to do so.

8.11 On hearing the fire alarm all customers and staff are to leave the premises via the designated fire exit routes. These are fire exit doors and corridors labelled with the sign 'Fire Exit'. Customers and staff must be directed to the assembly point, namely: *<insert assembly point>*.

8.12 Employees are expected to assist any disabled customers out of the building. Exit routes are on a level and obstructions should not deter exits.

Customers with hearing or sight impairments are to be guided to a place of safety.

In the event that a member of staff has a disability that may impede their means of escape from the building, they must have a Personal Emergency Evacuation Plan (PEEP) drawn up. Head Office personnel and the Training Department will assist with this. PEEPs must be followed for each employee, as applicable.

8.13 The Duty Manager will continue to liaise with the fire brigade and will notify senior managers of: *<insert name of business>* where necessary.

8.14 Once the situation has been brought under control and the premises deemed safe for re-occupancy (only on the instruction of the fire brigade) the Duty Manager will supervise the return of customers and staff.

8.15 The Duty Manager will conduct a review of the event and will record key findings, comments, etc. in the Fire Log Book or other suitable document for future discussion.

8.16 The Premises Manager will authorise a review of the fire risk assessment if deemed appropriate.

8.17 It is recognised that in the event of refurbishment work/works in progress certain means of escape may not be available for use. In such a case appropriate temporary measures will be implemented by the Premises Manager to ensure that the lives of persons present within the premises will not be put at risk. If necessary this will include reducing the safe occupancy level accordingly to compensate for the loss of means of escape, and the provision of additional fire-fighting equipment if hot works are being undertaken. In addition, if the alarm is required to be taken out of use to allow essential works to take place, an alternative means of raising the alarm will be implemented.

Risk Assessments: Questions and Answers
ISBN 978-0-7277-6076-0

ICE Publishing: All rights reserved
http://dx.doi.org/10.1680/raqa.60760.303

Chapter 18
Commercial security and business risks

What does a business need to consider in order to achieve commercial or business security?

A business must consider every possible event that could adversely affect the running of the business, and implement plans and procedures to mitigate the risks to the future viability of the business.

Listed companies must have a formal risk management strategy that sets out the steps that will be taken to manage those risks.

Risk reviews should not only consider financial matters but also:

- terrorism
- cyber threats
- natural disasters
- pandemics/epidemics of disease
- fraud
- product sabotage
- fire
- strikes and employment issues
- utility malfunctions (e.g. power cuts)
- supply-chain collapse
- civil disturbance.

Is a commercial security risk assessment the same as a business continuity plan?

No, not necessarily so. A commercial security risk assessment will set out a logical review of the threats to your business and the impact those threats will have on the business if they are realised.

A business continuity plan will set out the steps that a business needs to take to keep trading so that income is not lost and customers are not left without goods and services.

A commercial risk assessment and a business continuity plan will often go hand in hand, as the risk assessment will inform the business continuity plan.

What is a security or commercial risk assessment?

It is the process of identifying internal and external threats and vulnerabilities, identifying the likelihood of the event, defining the critical functions necessary to continue an organisation's operations, defining the controls in place to reduce exposure, and evaluating the costs.

What is business resilience?

Business resilience is about safeguarding your business, its people and assets. It should be part of your everyday management planning. If and when a business is faced with disaster, that preparation can help minimise the impact and help speed recovery. Thus, business resilience and planning should be regarded as a priority for any business, and is equally critical for small companies as for large organisations.

Every year, around 20% of all businesses across the UK face an event that is unplanned, unwanted and sometimes challenges their very survival. Approximately 80% of affected businesses will never fully recover. That threat may come as a result of fire or flood, theft or fraud, or even terrorist action; but no matter what the cause, businesses that successfully recover to thrive again are those that have prepared a business resilience plan. This means they have:

- assessed the likely impact on the business of significant and potentially damaging events
- planned their response in advance
- tested the effectiveness of the plan and revised it where needed
- invested time, thought and, where necessary, money in managing risk.

What is risk identification?

Identifying the risks that are relevant to your organisation is a key component of any security risk assessment. Any risks that are not identified at this stage will not be considered for further analysis, and as such will not be treated, potentially leaving your organisation vulnerable if those risks occur.

From a security perspective, there are three primary elements that should be examined to identify risks that are relevant to your organisation:

- Establish the *critical assets* of the organisation. This will allow you to focus on the assets or dependencies that are essential for the organisation to function and achieve its objectives.

■ Identify appropriate *sources of risk* (traditionally referred to as 'threats') in order to better understand the range of potential factors that may affect your critical assets.

■ Determine the *potential areas of impact* on the organisation if the sources of risk occur against your critical assets.

Security risks can, therefore, be identified as:

Critical assets that may be affected by *sources of risk* resulting in *potential impacts* to areas of the organisation.

What are critical assets?

All organisations have critical assets without which they would not be able to operate effectively or deliver their key services. By identifying these assets you will be able to apply the risk assessment process more rigorously to those elements that are of greater importance to the running of your operations.

You should be realistic when identifying your critical assets in order to prevent the risk assessment process becoming too cumbersome. You should also take into consideration any key dependencies that the organisation depends on but that are not directly under its control.

In general terms, you should be asking: 'Which assets and dependencies if lost or interrupted would significantly impact on the organisation's ability to deliver its key objectives?'

When identifying critical assets you should consider grouping elements under the headings of *people*, *information* and *physical assets*. Thought should also be given under these headings to identifying key dependencies.

What are sources of risk?

Traditionally referred to as 'threats' within a security context, you should seek to identify both internal and external potential sources of risk to your organisation.

When identifying sources of risk it is important to look for those that have the *potential* to affect the organisation. At this stage you do not need to identify the possible impacts.

For the purpose of this book, sources of risk are defined as internal and external threats that have the potential to affect the organisation's critical assets.

Sources of security risk can be looked at in terms of both the potential perpetrator and the method of attack.

Potential sources of risk

Arson	Cyber attack	Armed assault
Hijacking	Sabotage	Rioting and civil unrest
Surveillance	Suicide bomber	Car bomb
Chemical attack	Biological attack	Radiological attack
Piracy	Package bomb	Hostage taking

What methods should we use to consider risks to the business?

Once you have a clear picture of your business you can begin to identify the risks. Review your business plan and think about what you could not do without, and what type of incidents could impact on these areas. Ask yourself:

- When, where, why and how are risks likely to happen in your business?
- Are the risks internal or external?
- Who might be involved or affected if an incident happens?

The following are some useful techniques for identifying risks.

Ask 'what if?' questions

Thoroughly review your business plan and ask as many 'what if?' questions as you can. Ask yourself, what if:

- you lost power supply?
- you had no access to the internet?
- key documents were destroyed?
- your premises were damaged or you were unable to access all or part of the building?
- one of your best staff members quit?
- your suppliers went out of business?
- the area your business is in suffered from a natural disaster?
- the services you need, such as roads and communications, were closed?

Brainstorm

Brainstorming with different people, such as your accountant, financial adviser, staff, suppliers and other interested parties, will help you get many different perspectives on risks to your business.

Analyse other events

Think about other events that have, or could have, affected your business. What were the outcomes of those events? Could they happen again? Think about what possible future events could affect your business. Analyse the scenarios that might lead to an event and what the outcome could be. This will help you identify risks that might be external to your business.

Assess your processes
Use flow charts, checklists and inspections to assess your work processes. Identify each step in your processes and think about the associated risks. Ask yourself what could prevent each step from happening and how that would affect the rest of the process.

Consider the worst-case scenario
Thinking about the worst things that could happen to your business can help you deal with smaller risks. The worst-case scenario could be the result of several risks happening at once. For example, someone running a restaurant could lose power, which could then cause the food to spoil. If the restaurant owner was unaware of the power outage or the chef decided to serve the food anyway, customers could get food poisoning, and the restaurant could be liable and suffer from financial losses and negative publicity.

Once you have identified risks relating to your business, you will need to analyse their likelihood and consequences, and then come up with options for managing them.

How should we analyse and evaluate the impact of risks within our business?
Once you have identified the risks to your business you need to assess the possible impact of those risks. You need to separate minor risks that may be acceptable from major risks that must be managed immediately.

Analysing the level of risk
To analyse risks you have identified you need to work out the likelihood of an event happening (frequency or probability) and the consequences it would have (the impact). This is referred to as the 'level of risk', and can be calculated using the formula:

Level of risk = Consequence × Likelihood

Level of risk is often described as 'low', 'medium', 'high' or 'very high'. It should be analysed in relation to what you are currently doing to control the risk. Keep in mind that control measures decrease the level of risk but do not always eliminate it.

A risk analysis can be documented in a matrix as shown below.

Likelihood scale example

Level	Consequence	Description
4	Very likely	
3	Likely	
2	Unlikely	
1	Very unlikely	

Consequences scale example

Level	Consequence	Description
4	Severe	Financial losses greater than £50 000
3	High	Financial losses between £10 000 and £50 000
2	Moderate	Financial losses between £1000 and £10 000
1	Low	Financial losses less than £1000

Note: Ratings vary for different types of business. The scales above use four different levels, but you can use as many levels as you need. Also, use descriptors that suit your purpose (e.g. you might measure consequences in terms of human health rather than financial value).

Evaluating risks

Once you have established the level of risk, you then need to create a rating table for evaluating the risk. Evaluating a risk means making a decision about its severity and ways to manage it.

For example, you may decide the likelihood of a fire is 'unlikely' (a score of 2) but the consequences are 'severe' (a score of 4). Using the tables and formula above, a fire will have a risk rating of 8 (i.e. $2 \times 4 = 8$).

Risk rating table – example

Risk rating	Description	Action
12–16	Severe	Needs immediate corrective action
8–12	High	Needs corrective action within 1 month
4–8	Moderate	Needs corrective action within 3 months
1–4	Low	Does not currently require corrective action

Your risk evaluation should consider:

- the importance of the activity to your business
- the amount of control you have over the risk
- potential losses to your business
- any benefits or opportunities presented by the risk.

Once you have identified, analysed and evaluated your risks, you need to rank them in order of priority. You can then decide which methods you will use to treat unacceptable risks.

Commercial risk assessment is very similar to health and safety risk assessment (i.e. identify the hazard, determine who might be harmed, and how and how frequently, and then determine the control measures necessary to reduce or eliminate the risks).

I have heard of business impact assessments or analysis. What are these?

A business impact analysis identifies the activities in your business operations that are key to its survival. These are referred to as 'critical business activities'. You should consider things such as:

- the records and documents you need every day
- the resources and equipment you need to operate
- the access you need to your premises
- the skills and knowledge your staff have that you need to run your business
- external stakeholders you rely on or who rely on you
- the legal obligations you are required to meet
- the impact of ceasing to perform critical business activities
- how long your business can survive without performing these activities.

As part of your business impact analysis you should assign recovery-time objectives to each activity to help determine your basic recovery requirements. The recovery-time objective is the time from when an incident occurs to the time when the critical business activity must be fully operational in order to avoid damage to your business.

Your business impact analysis will help you develop a recovery plan, which will help you get your business running again if an incident does occur.

Key questions in a business impact analysis

To conduct a business impact analysis for your business, ask yourself:

- What are the daily activities conducted in each area of my business?
- What are the long-term or ongoing activities performed by each area of my business?
- What are the potential losses if these business activities could not be provided?
- How long could each business activity be unavailable for (either completely or partially) before my business would suffer?
- Do these activities depend on any outside services or products?
- How important are the activities to my business? For example, on a scale of 1 to 5 (1 being the most important and 5 being the least important), where would each activity fall in relation to the rest of the business?

As the risks to your business change, so too will their potential impacts. When you update your risk management plan, you will also need to conduct a new business impact analysis.

What should we do once we have identified the key commercial security risks to the business?

You will need to determine the business' appetite for risk.

First, double check that the board and senior management or business owner agree with your analysis of the business risks, and which people and tasks are essential. This will give you a clear understanding of the 'appetite for risk' within the organisation.

For example, one department may tell you it is essential to the business for them to be operational again within a day of any incident. It is up to the board to agree if it is willing to accept the risk of that department not being operational again within the agreed time or if it would rather plan to reduce the risks.

Next, define your strategy for dealing with risk. Whatever kind of business you are, you will probably choose one of the proven strategies. These are:

- Accept the risks – change nothing.
- Accept the risks, but make a mutual arrangement with another business or a business continuity supplier to ensure that you have help after an incident. That business could be a competitor.
- Attempt to reduce the risks.
- Attempt to reduce the risks and make arrangements for help after an incident.
- Reduce all risks to the point where you should not need outside help.

Are you the kind of business that is committed to reducing risks, or one that prefers to take risks and have a 'comeback' plan? Your management's attitude to risk may be partly based on the costs of delivering effective business continuity. When working out these costs, remember to include both money and people's time.

What is a security register and is it a good idea to have one?

A security register will contain all the important information regarding your business' future commercial security, and information on business risks and the steps you plan to take to mitigate them.

The register should contain the following:

- a copy of the assessment
- details of any security already in place
- guards
- alarm systems
- key holders
- a plan of the premises.

By using the plan of the building you will be able to identify the profile of each area and annotate the plan with key information such as access points, blind spots, critical storage areas, and areas vulnerable to public access or arson.

Regular checks will need to be carried out, and the register should contain a suitable place for all the records and checklists.

Actions taken and those outstanding should be included in the register.

A senior member of the company should be given the role of 'security liaison'. Any security lapses or problems should be logged and any problems resolved as soon as practicable. All staff should be included in highlighting problems and have some responsibility for the security of the company.

Where a company has external compounds or grounds, everything should be kept in its own area, (pallets, skips, stock, etc.), and every effort should be made to keep the outward impression of the company as one of efficiency and effectiveness.

The perimeter of the premises is most important, and damaged fencing should be remedied promptly. Broken windows should be replaced as quickly as possible, as failing to do so will generally lead to further incidents.

Where appropriate, all alarm activations should be kept in the register.

The list of company key holders should be kept up to date, and there should be a strict policy on keys and access.

All checks made by the insurance company or a crime reduction officer should also be recorded in the register.

What should be included in a business continuity plan?

A business continuity plan will set out the steps you will take when an emergency occurs that could stop your business from functioning. The plan should be a document that sets out what people are to do should all other information be inaccessible to them – it should be a document that you can pick up at home or in another office and know what to do to keep the business running and customers supplied with goods and services.

Typical contents of a manual will be:

- information on how to use the manual
- recommended procedures for updating the manual
- an overview of business continuity – key risks to the business
- the purpose of plan – why it has been produced
- the outcome of plan – what it is meant to achieve
- the objectives of the plan
- key staff (gold command) – who is to do what and when
- staff welfare and staff contact details
- how to communicate with staff
- how to communicate with customers and suppliers
- equipment critical to the business and procedures for replacing equipment
- suppliers critical to the business and how to ensure continued stock or services
- IT services and how to maintain contact with business-critical contacts, and how to ensure the integrity of the IT infrastructure
- premises – whether access will be denied
- replacement premises or alternative solutions

- financial procedures – maintaining banking, payment processes and salary payments
- insurance details and contact numbers
- procedures for testing, reviewing and updating the business continuity manual.

The manual should contain whatever information is critical to ensuring that the continuity of the business is maintained.

Businesses that are prepared are more likely to survive an emergency than those that have made no contingency plans.

Numerous organisations, including many local authorities, provide free templates for business continuity plans – use a web search to obtain a selection of templates and choose the one most suitable for your business.

Should we require our suppliers to have a business continuity plan?

It is definitely a good idea to discuss business continuity with your key suppliers, as their failure to provide you with the goods and services you need could cause you to go out of business.

Identify your key suppliers, including any software providers, and request a copy of their business continuity plan or, at the very least, written confirmation of how they will ensure that you will be provided with your vital goods and services in the event of an emergency. Have they considered how they will continue in business? What are their risks?

Do they have alternative sources of supplies should their key suppliers not be available? What will be the timescales? How will they let you know about any disruption to your supplies or services? Would your competitors be able to take advantage of any interruption in the supply chain?

It is increasingly vital to understand a key supplier's business and its threats and weaknesses. Together you may be able to develop a mutually beneficial plan to ensure that both businesses will continue to provide whatever is required for their respective customers and clients.

Review all purchase agreements and procurement specifications, and make sure that there is a section on requesting business continuity information so that you can be reassured that the supplier has continuity plans in place.

The Buncefield explosion and fire 2005 – Northgate Information Services

Had it not come so perilously close to tragedy, Northgate Information Services might have looked upon the Buncefield oil depot as the kind of marketing that money cannot buy. The software services provider and business process outsourcing partner would be hard pressed to demonstrate the reliability of its services any better than having withstood the near total destruction of its headquarters with little – relatively speaking – disturbance to delivery.

In December 2005, a small explosion at the Buncefield fuel depot near Hemel Hempstead triggered a huge subsequent detonation. The impact of the explosion, which mercifully occurred around 6 a.m. on a Sunday, when few people were nearby, measured 2.3 on the Richter scale, and the resulting fire was the largest fire in Europe since World War II.

Northgate's headquarters, situated directly next door to the depot, took the brunt of the blast, and several fires were ignited in the Northgate building. More than 200 production systems were destroyed, along with a newly refurbished call centre and the company's email hub.

Following a conference call early on Sunday morning, Northgate's disaster–recovery partner, SunGard, fired up its customer's back-up servers. By Christmas Day, all of Northgate's 212 internal systems were restored. Of the 40 terabytes of data that had to be recovered, only one tape's worth was lost.

This event was the ultimate test of Northgate's disaster recovery, business continuity and crisis management plans. The first task for the company's emergency management team was to prioritise the services that had to get back online immediately, explained Mark Farrington, who was at that time Northgate's managing director of public sector and corporate services, and is now its recovery director.

'In our HR services department, which manages the payroll for 30% of UK workers, there is a rolling pay-out window, because different companies pay their staff at different times,' said Mr Farrington. 'A business continuity plan is useless if it fails to take account of the time of month or year that the hypothetical disaster occurs.'

Once systems had been prioritised, Northgate's technical staff of over 100 began to work on getting systems back on line, putting in 12-hour shifts back to back.

'Northgate, being a technology company, was lucky to have these resources available,' said Mr Farrington, 'but had we lost any of the 30 core support staff that knew the systems best, we would have been stuck.'

Also fortunate was Northgate's ownership of a recently deserted building in nearby Dunstable, which became an operational site. This helped maintain a sense of purpose, and even community, for employees who, under many other companies' business continuity plans, would have been working at home. 'It was important to ensure that employees didn't feel alienated from the organisation,' said Mr Farrington. 'In a disaster, interaction is key.'

Critical to managing the recovery effort was a flexible communications system, he said. 'It was important to establish a staff communication hotline on day one, so that people knew where they stood. Both our extranet and the SMS messaging service we have were key to getting information out there.'

New phones had to be issued to key employees, after their mobiles became clogged with concerned calls from friends and relatives.

Although Northgate's recovery of services was rapid, Mr Farrington warned against thinking that 'business as usual' can be returned immediately. 'You have to maintain focus on restoring the business to full capacity. Ours is a business that took 30 years to build. That's not going to come back online in 10 days.'

(Source: Information Age)

SECURITY – CHECKLIST

This checklist may help you to identify the areas in your business where you may be vulnerable. It is not designed to cover all aspects of security, but it will identify some common vulnerabilities.

Visitor access to the building
Are visitors allowed entry to your building by appointment only?

Do they have to report to a reception area before entry?

Are visitors asked for proof of ID?

Are they provided with visitors' badges?

Are all visitors asked to sign in when they enter the building?

Are visitors' badges designed to look different from staff badges?

Are all visitors' badges collected from visitors when they leave the building?

Does a member of staff accompany visitors at all times while in the building?

Are the returned visitors' badges cross-checked against the list of those issued?

Do your staff wear ID badges at all times when in the building?

Physical security of the building
Are there good-quality locks on all doors and windows at ground level?

Are there good-quality locks on each accessible door and window above ground level?

Can internal doors be locked when left unattended for long periods?

Are all fire doors alarmed?

Do you nominate members of staff to check that all doors and windows are closed and locked at the end of the business day?

If you have a burglar alarm, are your staff familiar with the procedures for switching it on and off? (In order to reduce false alarms.)

Do you maintain good visibility around the perimeter of your building (e.g. cutting back overgrown planting)?

Do you have adequate lighting around your building during the hours of darkness?

CCTV

Are your CCTV cameras regularly maintained?

Do the CCTV cameras cover the entrances and exits to your building?

Do you have CCTV cameras covering critical areas in your business, such as server rooms or cash offices?

Do you store the CCTV images in accordance with any guidance issued by the local police?

Could you positively identify an individual from the recorded images on your CCTV system?

Information security

Do you lock away all business documents at the close of the business day?

Do you have a clear-desk policy out of business hours?

Do you close down all computers at the close of the business day?

Are all your computers password protected?

Do you have firewall and antivirus software installed on your computers?

Do you regularly update this protection?

Do you employ the principle of least privilege?

Do you back-up business-critical information regularly?

Security checking of personnel

It is important to prove the identity of potential new staff. You should see original documents, not photocopies, and, where possible, check the information and request an explanation for any gaps. During the recruitment process, do you ask for:

Full name?

Current address and any previous addresses in the last 5 years?

Date of birth?

National insurance number?

Full details of references (names, addresses and contact details)?

Full details of previous employers, including dates of employment?

Proof of relevant educational and professional qualifications?

Proof of permission to work in the UK for non-British or non-European Economic Area (EEA) nationals?

Full (current) 10-year passport?

British driving licence (ideally the photo licence)?

Birth certificate – issued within 6 weeks of birth?

Credit card, with three statements and proof of signature?

Cheque book and bank card, with three statements and proof of signature?

Proof of residence (council tax, gas, electric, water or telephone bill)?

Full EEA passport?

National identity card?

A Home Office document confirming the individual's UK immigration status and permission to work in the UK?

Communication
Do you have a security policy or other documentation showing how security procedures should operate within your business?

Is this documentation regularly reviewed and, if necessary, updated?

Do you have a senior manager who takes responsibility for security within your business?

Do you regularly meet with staff and discuss security issues?

Do you encourage staff to raise their concerns about security?

Are you a member of a local Business Watch or a similarly constituted group?

Do you know your local community police officer or community support officer?

Do you speak with neighbouring businesses about issues of security and crime that might affect you all?

Do you remind your staff to be vigilant when travelling to and from work, and to report anything suspicious to the relevant authorities or police?

What do your results show?
Having completed the checklist, you need to give further attention to the questions that you have answered 'no' or 'don't know' to.

If you answered 'don't know' to a question, find out more about that particular issue to determine whether that vulnerability is being addressed or needs to be addressed.

If you answered 'no' to any question then you need to address that particular vulnerability as soon as possible.

Where you have answered 'yes' to a question, remember to regularly review your security needs to make sure that your security measures are fit for purpose.

BUSINESS CONTINUITY RISKS

Which of the following hazards and threats are relevant to your business, department or service?

	Hazard or threat	Yes or no?	Why?
1	Fire or flood to the building or surrounding areas		
2	Loss of electronic records, especially of customers' contact details		
3	Loss of paper records including financial information, banking records, invoicing, etc.		
4	IT systems/application failure		
5	Mobile telephone failure leading to inability to communicate with staff and customers		
6	Major IT network outage, failure of service partner (e.g. payroll provider, website hosting company		
7	Denial of access to premises or surrounding areas (e.g. main roads and common office areas)		
8	Terrorist attack or threat affecting the transport network or office locations		
9	Theft or criminal damage of equipment, infrastructure, records, cash, etc.		
10	Chemical contamination in the area, major chemical spillages, terrorist attacks		
11	Serious injury to, or death of, staff while in the offices, causing major investigations and enforcement actions		

Hazard or threat		Yes or no?	Why?
12	Significant staff absence due to severe weather or transport issues or due to inaccessibility to office accommodation		
13	Infectious disease outbreak among staff members, causing widespread absenteeism		
14	Simultaneous resignation or loss of key staff		
15	Industrial action caused by strikes or actions in adjoining premises		
16	Fraud, sabotage or other malicious acts, especially of company funds, products		
17	Violence against staff		
18	*Please add any other relevant threat*		

BUSINESS IMPACT ANALYSIS (BIA)

Complete the tables for each major function of your business (e.g. access to building, availability of key staff, suppliers, IT, sales and marketing, and finance).

Each table details, over time, the impact the loss of this function would have on your ability to meet your organisation's aims and objectives and the impact on your stakeholders, including staff, investors and suppliers.

Key function/service:		
Time	Impact description (Minor, moderate or significant)	Example of impact to service
First 24 hours		
24–48 hours		
Up to 1 week		
Up to 2 weeks		

Consider the maximum length of time that you can manage a disruption to this key function/service without it threatening your organisation's viability. At what point in time would you need to resume this function/service in the event of disruption? This is called the recovery time objective (RTO).

Resources required to meet the recovery time objective

Recovery time objective:					
Time	People	Premises	Technology	Information	Suppliers/Partners
First 24 hours					
24–48 hours					
Up to 1 week					
Up to 2 weeks					

Key business impacts

Loss of systems (IT or telecommunications)

Loss of utilities (e.g. water, gas or electricity)

Loss of or access to premises

Loss of key suppliers

Disruption to transport

Major reduction in staff numbers (e.g. due to pandemic disease)

Loss of key Directors or board members

Risk Assessments: Questions and Answers
ISBN 978-0-7277-6076-0

ICE Publishing: All rights reserved
http://dx.doi.org/10.1680/raqa.60760.325

Index